Pvt. Ltd.

Plant Genetic Engineering
— *Elements* —

Debashis Banerji

Professor Emeritus
Baba Amte Centre For People's Empowerment
Samaj Pragati Sahayog, Bagli, District Dewas - 455227
(*Formerly*, Professor & Head, Botany Department,
Institute of Advanced Studies, Meerut University, Meerut)

2013

Studium Press (India) Pvt. Ltd.

Plant Genetic Engineering
— *Elements* —

ISBN: 978-93-80012-61-2

Published by:

Studium Press (India) Pvt. Ltd.
4735/22, 2nd Floor, Prakash Deep Building (Near PNB),
Ansari Road, Darya Ganj, New Delhi-110002
Phone: +91-11-43240200-15 (15 lines); Fax: +91-11-43240215
jngovil@gmail.com; jngovil@hotmail.com

Printed at: **Salasar Imaging Systems**

ABOUT THE AUTHOR

Dr. Debashis Banerji has been a student of Plant Physiology and Molecular Biology since 1960. From early 1960s to 1993, he has researched on and taught these subjects at the Botany Department, Allahabad University, Allahabad; Orchard Park Labs of Roswell Park Memorial Cancer Research Institute (RPMI), Buffalo, New York; Trocaire College, Buffalo, NY and Botany Department, Meerut University, Meerut. He got his doctorate in 1966 from Allahabad University, on the role of cytokinins in plant development. At RPMI, Buffalo, NY he did pioneering post doctoral work on plant histones and chromatin associated nucleases in a lab which, in 1968, was one of the first to show *Agrobacterium* DNA integration into tobacco DNA for induction of crown gall tumour. At Meerut University, over thirty students have done doctoral work with him on areas related to plant development and heavy metal pollution. The researches have resulted in some valuable publications in high impact journals including Nature, Plant Physiology and BBRC. In 1993 he joined Samaj Pragati Sahayog (SPS), a voluntary organization of young visionaries, inspired by Baba Amte and dedicated to rural development through the Watershed Development approach and people's empowerment. At SPS he has been involved with development and spread of location specific sustainable dryland agriculture packages and empowerment of farmers and village workers through its Baba Amte Centre for People's Empowerment. From mid nineties he has been specially involved with theoretical and on-field aspects of GM crops, specially Bt cotton. Many scientists have helped him in this endeavour. The book Plant Genetic Engineering-Elements is based on these experiences of the author.

PREFACE

From early seventies, Recombinant DNA science has expanded the understanding of genes and genomes to extents, inconceivable even in late sixties. Besides addition to basic information, the greatest achievements of recombinant DNA technology or Genetic Engineering have been in its applications for human good. The possbility of gene transfers across species, enabled by this technology, madc it possiblc to multiply and study any gene or a DNA sequence in a suitable host as a bacterium, like *Escherichia coli*. The learnings from such studies, of chromosomes, genes and genome sequences have been applied in areas as, gene therapy, production of pharmaceuticals and vaccines and specially, crop improvement.

The author has been a student of Plant Physiology and Molecular Biology since 1960 and was lucky to be associated with a laboratory in the US (Orchard Park Labs, Roswell Park Memorial Cancer Research Institute, Buffalo, New York) which was one of the first to indicate, in 1968, the integration of *Agrobacterium tumefaciens* DNA into tobacco DNA for induction of crown gall tumour. On return to India, one could continue the learning, at least mainly theoretically, at the Institute of Advanced Studies, Meerut University, Meerut (now CCS University), thanks to the academic support of, Late Professors V. Puri and Y. S. Murty.

The author was specially lucky to be able to have, since 1993, hands on learning of the value of crop improvements introduced by agriculture scientists. This happened during his field work as a member of the voluntary organization Samaj Pragati Shayog, which works in close contact with Agriculture Universities and ICAR

Institutes of mainly MP and ICRISAT. The lab and land experiences have helped in developing better insights into the value of genetic modifications.

The present book is an attempt to introduce the students to the basic elementary aspects of the subject with emphasis on plant genetic engineering. It is hoped that this book will help the students to understand and enjoy more advanced treatises written by scientists in India and abroad.

Debashis Banerji

ACKNOWLEDGEMENTS

I am grateful to my under graduate and post graduate students and research coworkers at the Universities of Allahabad and Meerut, whose queries in the class rooms and labs have helped me in understanding the subject better. My heartfelt thanks to Professor M. M. Lalorya, my teacher and doctoral research guide for initiating me into the subject and to Professors Govindjee (University of Illinois, Urbana, USA), S. P. Sen (Kalyani University, WB) and V. S. Ramadas (S.V. University, Tirupati) for being sources of inspiration.

My special thanks to Professors Shanti Sharma, and Shashi Sharma, Bioscience Deptt, H. P. University, Shimla; Professor Vimala Y, Botany Deptt, CCS University, Meerut and Dr. V. S. Reddy, ICGEB, New Delhi, for perusing some of the chapters and giving their inputs. Thanks to Navin Digitech of Indore for transforming my rough sketches to good diagrams.

I am grateful to Rita Banerji, for providing a peaceful space and ambience for writing some of the chapters in Delhi.

My real life lab to land transfer took place when I joined, Samaj Pragati Sahayog (SPS), in 1993. SPS is a voluntary organization dedicated to holistic rural development, in a remote dryland tribal area of Disrict Dewas, MP inspired by Baba Amte. Two decades of field work with hundreds of farmers growing many varieties of kharif (rainy season) and rabi (winter) crops have given a broad spectrum experience of agriculture. One witnessed, farmer's struggles against abiotic and biotic stresses and the endeavours of agriculture scientists in overcoming the hurdles through crop improvements and better practices. These crop improvements combined with the SPS approach of holistic Watershed Development, involving, farmland productivity enhancement and employment generation, have brought prosperity in the SPS areas of work. These experiences have convinced the author about the needs for new technologies and spread of information of technologies to farmers, their empowerment, for a successful participatory approach towards better farming in the country. I am extremely grateful to the young, visionary and dedicated founder members of Samaj Pragati Sahayog, specially, Dr.Mihir Shah, now member Planning Commission, for giving me this opportunity and also providing me, in the last two years, all the facilities and liberty for writing this book.

I could keep up with the subject, despite field work in a remote rural area, because of support and encouragement of mainly Drs PK Gupta, V Singh and AK Srivastava (Meerut University), S Bharti, KN Guruprasad, S Patil and S Chand (DAVV, Indore), VK Rai (H P University, Shimla), V Sharma (Vanasthali University), R M Agarwal (Jiwaji University, Gwalior), N Kumar (NBRI, Lucknow), U. Lavania (CIMAP, Lucknow), S Leelavathi (ICGEB, New Delhi), AK Tyagi (Director, NIPGR, New Delhi), and my friends late Drs. VK Laloraya (CMP Degee College, Allahabad) CM Govil and VK Gera (Meerut University).

My pranams to Paramhans Swami Niranjan of BSY, Monghyr, Bihar.

My heartfelt thanks to Dr. JN Govil of Studium Publishers for his guidance, patience and encouragement during the preparation of this book.

I thank my late parents for their blessings, all my children and grand children for being my anchors. I specially thank God for his gift of the divine presence of Mridula by my side, all along.

CONTENTS

1

Introduction

INTRODUCTION

Recombinant DNA molecules are DNA sequences that result from the use of molecular cloning methods, to join together genetic material from multiple sources. Such sequences are not otherwise found in biological organisms. Recombinant DNA molecules are sometimes called chimeric DNA, because they are usually made of material from two different species, like the mythological *chimera*. Unique recombinant DNA sequences may be created and introduced into any of a very wide range of living organisms. It is important to note that recombinant DNA differs from genetic recombination in that the former results from artificial methods in the test tube, while the latter is a normal biological process that results in the remixing of existing DNA in a species. This makes *recombinant DNA technology* or *genetic engineering* a landmark advance in biotechnology. It is now possible to cut a gene from any species and paste it into the genome of another species. This is the basis of genetic engineering of microbes, animals and plants and the production of Genetically Modified Organisms or *GMOs*. It is also possible to isolate a gene or any DNA segment from any species, introduce it into a bacterium like *Escherichia coli*, multiply it and study its structure and function, the growing area of *functional genomics*. The same strategy of inserting a gene in a microrganism like, *E. coli* or yeast and engineering its expression, is being used for manufacturing many biomaterials as vaccines. The aim of this book is to introduce the students, to the basic aspects of this science with some emphasis on plant genetic engineering. It is hoped that this will enable them to explore the subject further, through more advanced literature available in the form of texts and websites.

Chapter 2 gives in brief, a historical perspective of how the advances in life science and the developments in agriculture and biotechnology have been interlinked. For instance, the traditional art of fermentation led finally to the establishment of the cell theory and later to the discovery of enzymes. Similarly, the landmark findings of Gregor Mendel, in mid nineteenth century, of dominant and the independently assorted traits have gone a long way in improvements of crops, ranging from the use of native genes to recombinant genes. At the same time it has been pointed out that, for improvements in agriculture a holistic approach is essential. Crop improvement programmes must be coupled with programmes of regeneration of the environment and employment generation.

Understandings of cell and molecular biology constitute the backbone of biotechnology, specially genetic engineering. Chapter 3 starts with a description of the basics of the structural organization of cells and some macromolecules. Further, fundamentals of replication, gene expression and its regulation in prokaryotes and eukaryotes have also been described.

The experimental path is the grammar of any science. It is good for the students to have a proper background of the techniques and experimental designs to appreciate scientific discoveries. This becomes more important in life science, as, living organisms are dynamic systems. With this in view, in Chapter 4, some important techniques have been described. The methodologies have been described to illustrate how a gene DNA segment is finally isolated, ligated to a vector to form a recombinant DNA, is amplified by cloning or polymerase chain reaction, for subsequent structural and functional analysis.

The key to success of recombinant DNA technology, is the progress in the development of the carrier DNAs or Vectors. It is because of ligation with vector molecules that a gene or DNA of interest from any species can be amplified, expressed and studied in model bacterial or eukaryotic systems. It is the vector molecules which can jump species barriers. The main development has been in the carrying capacity of the vector and host range potential. In Chapter 5 simple descriptions of the designing of different types of vectors have been given. Cloning and the expression vectors described, include vectors derived from bacterial plasmids, phage and yeast for cloning in prokaryotic and eukaryotic cells.

All students need libraries of books. Similarly, scientists researching on genomes, need a collection of genes and DNA sequences, in good amounts, for investigations. Such pools of DNA fragments have been created mainly by cloning. A collection of such clones of DNA is termed as a *genomic library*. However, in a regular library, the books are catalogued and labeled and so it is easy to find a book of interest. But, in a genomic library it is all unknown. The clones have to be screened to detect a gene of interest. Chapter 6, therefore, gives an introductory account of DNA libraries. The topics covered include, genomic library, cDNA library, as also some screening methods.

According to the central dogma of molecular biology, DNA gives rise to RNA and RNA to proteins. This is the main basis of the expression of traits in an organism. Fundamental to this phenomenon is, the sequence complementarity between DNA and RNA and colinearity between protein sequences and those of the nucleic acids. Therefore, scientists have been trying to unravel the sequences of these macromolecules since long. Such endeavours ultimately led to the Human Genome Project. HGP has been a remarkable example of concerted scientific venture at global level. This programme of human genome sequencing catalysed the sequencing of many microbial, animal and plant genomes. In Chapter 7 some basic aspects of HGP as, genome components, genome sequencing strategies, chromosome mapping, DNA sequence markers, whole genome sequencing and related aspects have been described. At the end, some milestones in the history of human genome project have also been mentioned.

Chapter 8 is devoted to transgenic plant production. The production involves a two step method of at first, production of the *Primary Transformant* by genetic engineering, and then the *GM crop for release* to farmers, by transgene transfer from the primary transformant to a crop variety by conventional plant breeding. Further, brief accounts have also been given of some marker genes in plant transformation, marker free transgenic crops, promoters, plant tissue culture, plant breeding basics, transgenes and transgenic crops, testing of GM crops by transgene analysis and agronomic performance. Some emphasis has been laid on the importance of the enabling techniques. A cell or tissue may be transformed, but, if it cannot be regenerated by some tissue culture process, the whole effort goes waste. Therefore, tissue culture is a very important and essential technique. Moreover, in attempts to highlight genetic engineering technology, conventional breeding is sometimes sidelined. Nothing can be farther from truth. The students must realize that, real life lab to land transfer of GM crops cannot happen without conventional plant breeding. It has been observed that not all varieties of a species can accept and integrate an alien transgene. Usually, the cells or tissues which can be transformed are, from lines/ varieties of a species which are not agriculturally suitable, for instance, having very low yields. Therefore, the transgene has to be transferred from the primary transformant to an agriculturally suitable variety by a proper protocol of conventional breeding, as backcross. It needs a committed team of agriculture scientists as, breeders, agronomists, entomologists and others to ensure stable performance of a GM crop, through tedious pre-release investigations. A lot of pre release investigations are also necessary for assessment of biosafety with regard to impacts on specially, human health and the environment. Thus, GM crop production is a concerted effort of recombinant DNA scientists and scientists from the areas majorly of, tissue culture, agriculture, health and environment.

So far, the commercially released G M crops have been made by nuclear transformation. In contrast to the nucleus, cells have many chloroplasts. Besides difference in numbers, a major difference is that, the nuclei are inherited through both the pollen and egg in the higher plants while chloroplasts are passed on to the progeny from the egg cells alone. Inheritance of nuclei through pollen by pollination is paternal inheritance and inheritance of nuclei through the egg is maternal inheritance. Therefore, chloroplast inheritance through egg is also maternal inheritance. The egg rests in the ovary, has no chance to come in contact with reproductive parts of other plants and therefore, all the genes, its own and those derived from the pollen after fertilization, pass on only to the progeny embryo after fertilization. *So, there is no chance of flow of maternally inherited genes to other plants.* Pollens are however, let loose, float around and can fall on the stigma of the flowers of other plants too and may induce pollination in related species. Thus, *paternally inherited genes have the potential to flow to other species.* Therefore, plants modified by nuclear transformation have the possibility of spreading the transgene and marker genes through horizontal gene transfer to other plants by pollination. The chloroplast genes are maternally inherited and do not flow to other plants and so, a transgene introduced in the chloroplasts remains contained. Further, a nucleus will have one copy of a transgene in a cell, whereas if introduced in chloroplasts, there will be many copies of the transgene per cell. Therefore, transgene product of GM crops generated by chloroplast transformation will be manifolds more than the GM crops produced by nuclear transformation. Scientists engaged in genetic engineering have been concerned about the possibilities of gene flow and have endeavoured to design biosafe and better GM crops. With this aim, the transplastomic plants, GM crops through chloroplast transformation, were developed. In these plants, there is maximum gene containment and the amounts of transgene encoded protein is high. Chapter 9 covers some aspects of the transplastomic plants. The topics included are, the plastid genome, chloroplast transformation, applications of chloroplast transformation in functional genomics, transplastomic plants of economic importance.

Chapter 10 deals with Intellectual Property Rights and Biosafety Regulations. Production of GMOs initiated the era of patenting novel living organisms. The US Supreme Court, in the famous Diamond vs Chakravarty case, for the first time, allowed such patenting. The implications of such patents and types of intellectual rights have been dealt with briefly. Biosafety is a major concern. Scientists around the world have introduced innovations in the technology for production of biosafe GM crops. Such measures include production of marker free GM crops, use of safe tissue specific and inducible promoters and even *cisgenic* plants with plant based promoters. However, mainly and possibly due to prohibitive costs, these innovations have not been incorporated in the commercial production protocols so far. Hopefully, some solutions will come for the way ahead.

In most of the chapters, the author has tried to give mainly the facts and the findings. Historical aspects have been avoided. Therefore, at the end, in chapter 11, is given a compilation of mile stone findings in cell and molecular biology.

It may be mentioned that the author has tried to maintain a class room language, in the descriptions, as far as possible. The figures are also self drawn with the aim that the reader may also easily draw them if necessary.

2

Overview of Developments in Biotechnology and Genetic Modification

SUMMARY OF CONTENTS

This chapter gives in brief, a historical perspective of how the advances in life science and the developments in agriculture and biotechnology have been interlinked.

Biotechnology had been initiated thousands of years ago, specially when the roaming, hunting and food gathering nomadic human communities preferred to settle in suitable locations. Such communities began an agrarian livelihood and started domesticating plants and animals. That was the inception of biotechnology. Biotechnology involves use of living organisms or their materials to make some product/s and includes introduction of improvements in plants, animals and microrganisms for direct or indirect benefit to humans. Recombinant DNA technology or Genetic Engineering is a landmark turn in the history of biotechnology. In the following sections we look at the gradual development in the trends and ramifications in biotechnology.

Various archaeological studies indicate collection of wild type seeds, tubers, cuttings of plants with desired traits, their field screening and also preservation of such plant parts for future by early human communities. Similar practices were followed in the case of domestic animals as goat, sheep and cattle. Thus, the farmers were saving genetic stocks from early days. This was the ancient form of present day *germplasm conservation*. Such activities must have been happening initially at the centres of origin of various edible plants as identified by the great Russian scientist Nikolai Vavilov. The centres of origin are, I) China, II) South/Southeast Asia, III) Central Asia, IV) Middle East, V) Mediterranean, VI) East Africa, VII) Middle America, VIII) South America. Most of these centres were in the developing world. Encroachments into wild and traditional agriculture land due to rapid development interventions in general and expansion of modern agriculture has gradually been eroding agro-diversity. In view of this, from early 1970s gene banks have been developed through a global initiative, the CGIAR (Consultative Group on International Agriculture Research) which has seventeen main centres around the world including ICRISAT in India.

Archaeological studies also indicate early origins of *fermented foods and fermented beverages*. Thus, around 5000 to 4000 years ago, the use of fermentation (Latin, *fervere* = to boil) started, for making of food and beverages as bread, yogurt, wine and beer. Fermentation of sugary material led to formation of froth (appearance of 'boiling' due to evolution of bubbles of CO_2 and later alcohol formation. Similarly, fermentation of a dough of crushed wheat powder, flour, led to some aroma and rising of the dough due to the trapped carbondioxide. While bread and beer making seemed to have started in the Egyptian area, origin of yogurt was in the fareast Chinese region. The discovery of fermentation must have started with detection of useful effects of chance infection of raw materials as dough and sugary syrups by microbes. The method was regularized by the communities as an art, with the use of small amounts of fermented dough or fermented beverage from previous stocks as starters. Much later in late 1860s, the famous French scientist Louis Pasteur established that the microbe yeast is responsible for fermentation. Similarly, lactic acid bacteria were shown to be the agents for curdling of milk. The innovations in fermentation by the ancient communities are the progenitors of modern day *industrial biofermenters*. From 1900 onwards *biofermentation* processes have been used to produce many compounds of value as, antibiotics, primary metabolites as amino acids, enzymes, pharmaceuticals and other valuable compounds.

As mentioned above, *plant breeding* has been practiced for thousands of years, since nearly the beginning of human civilization. It is now practiced mainly by professional plant breeders of government research institutes, universities and private research centres. Breeding new crops is important for ensuring food security. The traits that breeders have tried to incorporate into crop plants include, increased yield and quality, increased tolerance to biotic and abiotic stresses, specially, resistance to drought, microbes and insect pests. Almost all the domesticated plants used today for food and agriculture were domesticated in their *centres of origin*. The great diversity of closely related wild relatives of crops in such centres can be used for improving modern cultivars by plant breeding even now. A crop plant whose origin or selection is due primarily to intentional human activity is called a *cultigen*, and a cultivated crop that has evolved from wild populations due to selective pressures from traditional farmers is called a *landrace*. Landraces, which can be the result of natural forces or domestication, are plants that are ideally suited to a particular region or environment. Examples are the landraces of rice, *Oryza sativa* subspecies *indica*, which was developed in South Asia, and *Oryza sativa* subspecies *japonica*, which was developed in China.

The detection of the usefulness of *heterosis* for plant breeding has led to the development of inbred lines that reveal a heterotic yield advantage when they are crossed. Further, the discovery of *cytoplasmic male sterility*, CMS, a maternally inherited trait that makes the plant produce sterile pollen enabled the production of hybrids without the need for labour intensive detasseling.

Nazareno Strampelli, in Italy produced a number of wheat hybrids (1904 to late 30s) which helped Italy to increase hugely crop production during the 'battle for grain' (1925–1940) and some varieties were exported to countries, as Argentina, Mexico, China and others. After the World war II, the work of Strampelli was nearly forgotten, but thanks to the hybrids he created, Norman Borlaug could initiate *Green Revolution* in the developing countries including India.

The development of modern biotechnology, through the years, is related to the progress in the understanding of cell structure, function, division, differentiation and reproduction at the microscopic and cellular levels. In 1660s Robert Hooke on the basis of microscopic observations of cork sections first described tissues to be composed of small compartments, *cellulae*, the cells. A little later, in 1670s Anton van Leewenhoek, a Dutch pioneer of microscopy also reported *animalcules*, microscopic organisms, in pond water droplets. It nearly took two hundred years for the German scientists Matthias Schleiden a botanist, and Theodore Schwann a physiologist to enunciate the *cell theory* in 1830s, which was further elaborated by the German pathologist Rudolf Virchow in 1850s. According to this theory, all organisms consist of cells and cells give rise to cells. Louis Pasteur, in 1860s, gave evidence for this concept by showing convincingly that microorganisms give rise to their progeny microorganisms and that the microbes do not arise *de novo* as believed in the earlier postulates of spontaneous generation of life. Pasteur, however, had believed in the vital theory, that conversion of sugar to alcohol requires living yeast cell. In 1890s, Eduard Buchner showed that, even yeast extracts could achieve this conversion and described the active principle as *enzyme* (in yeast). Enzymes were found to be proteins. Such findings led to intensive studies on the composition, structure and function of proteins and also various metabolic pathways. A finding of major interest was the fact that every protein had a specific amino acid sequence. How the cellular machinery achieved this sequence specificity for hundreds of proteins during their synthesis was intriguing and the answers came later.

Development of techniques and equipment as centrifuges and ultracentrifugation, tracer techniques, electrophoresis and other analytical tools were of great help in exploring the biochemistry and physiology of functional processes at organismal, cellular, organelle and molecular levels. Meanwhile the invention of electron microscope by the German engineer Ernst Ruska in 1930s and its refinements led to understanding of cell ultrastructures and helped investigators in correlating molecular processes with cellular organization.

Cells giving rise to similar cells and organisms begetting similar organisms raised the curiosity of how the fixity of traits is maintained during reproduction from generation to generation. What controls the traits, the characteristic features of an organism and how these traits pass from parents to progeny ? For ages, scientists have been seeking answers to these questions. The finding that in higher organisms, fertilization consists of fusion of male and female nuclei had indicated that the traits pass from parents to offsprings through the nuclei.

Around the time Pasteur was investigating fermentation, an Austrian monk and botanist Gregor Mendel was studying inheritance of contrasting traits in pea. Mendel suggested that traits are passed on from parents to offsprings as *units of heredity,* one unit for each trait. His discoveries published in 1865 were rediscovered in 1900 and came to be established as Mendelian Genetics. It was clear that these units of heredity existed in the nuclei. To find out the chemical which could be the possible chemical carrier of traits, a Swiss biochemist, Johann Freidrich Miescher, in 1869, isolated a high molecular weight, nitrogenous, phosphate rich acidic chemical from the nuclei of white blood cells and termed it nuclein, later named as *nucleic acid*. It must be appreciated that at Meischer's time, centrifugation did not exist. He got a sediment of nuclei, by suspending partly degraded WBC from pus in dilute acid which dissolved the cytoplasmic material surrounding the nuclei. In 1882, Walter Flemming of Germany, observed threadlike bodies in dividing cells which got copied and passed on to the daughter cells. As these structures became distinctly coloured with differential stains, they were termed as *chromosomes* (coloured bodies) by W. Waldeyer in 1888. In 1903 in America, Walter Sutton showed that the units of heredity occur in chromosomes and Wilhelm Johannsen, a Danish botanist, termed these units as *genes* in 1909. Years of studies on nucleic acid structure and function, described in the next chapter, established DNA as the undebatable chemical basis of heredity in most organisms. Several scientists as George W. Beadle, Edward L. Tatum and Charles Yanofsky demonstrated correlations between changes in the activities of certain enzymes and mutations of specific genes. This led to the one gene one enzyme hypothesis in 1950s. The enunciation of the double helical model of DNA by James D. Watson and Francis Crick in 1953, studies on DNA, RNA and protein synthesis, initial sequencing studies and manipulations with oligonucleotides established the main tenets of the *central dogma,* DNA gives rise to RNA and RNA to proteins. Although the basic tenets of the central dogma are true, certain exceptions exist. Thus, ribosomal and transfer RNA genes give rise to the respective RNAs only. Further, an astounding milestone of molecular biology has been the discovery of *reverse transcriptase* by Temin in late sixties. This enzyme was shown to make DNA from RNA and is widely used in making cDNA libraries. Jacob and Monod's concept of the *operon model* of prokaryotic gene expression and regulation, catalysed similar studies on eukaryotes since 1960s. These studies and specially the discovery of enzymes for DNA replication, repair, specific cutting by restriction enzymes and ligation of cut DNA fragments helped in DNA manipulations in vitro, paving the way for Recombinant DNA technology or Genetic Engineering. Now a gene of interest could be cut away from its host genome, joined to some other DNA fragment and studied independently in a test tube, in vitro. Such studies were initiated in early 1970s at the Universities of Stanford and San Francisco. The collaborative work of teams of Paul Berg, Stanley Cohen and Herbert Boyer laid the foundation protocols of the technology. The steps involved i) isolation of a gene by cutting with a restriction enzyme, b) construction of a *recombinant DNA* by joining it to a similarly cut DNA segment which could act as a vehicle or *vector* to carry the gene to a bacterial cell, c) introduce the gene-vector recombinant DNA to a bacterial cell ie *transform* a bacterial cell and d) selecting the transformed bacteria and e) studying the activities of the gene in the new host, f) if necessary isolate the much amplified recombinant DNA from the transformed bacterial colonies. The genes studied initially were antibiotic resistance genes, as, tetracycline resistance gene, isolated by cutting the source DNA with a restriction enzyme. The initial vector DNAs consisted of a bacterial plasmid DNA pSC101 and its derivatives. The extrachromosomal plasmids were known to move across bacterial cells and thus confer new traits, as antibiotic resistance, to the host. *As plasmids could move across cells and transfer genes, they were used as vectors.* The DNA containing the gene and the vector were cut with the same restriction enzyme. The products of same restriction enzyme have complementary ends. Thus, the cut ends of the gene and the vector being complementary, the two DNAs anneal and then are ligated leading to the formation of a *recombinant DNA*. The procedure of cutting a gene or DNA segment of interest, sticking it to a similarly cut vector to make a recombinant DNA and amplification through replication of the recombinant DNA in a host, like *E. coli* is known as *DNA cloning*.

Soon, the above scientists further demonstrated that by this protocol *genes could be transferred across species.* Thus they demonstrated transfer of different genes from the bacterium *Salmonella*, the frog *Xenopus laevis* and mouse, to *E. coli.* This was a quantum leap in the advances in biotechnology.

Applications of *Recombinant DNA Technology* or *Genetic Engineering*, introduced landmark advances in agriculture. An useful tool has been *Marker Assisted Selection*. This method uses molecular markers and DNA fingerprinting to map thousands of genes. The screening is based on the presence or absence of a certain gene as determined by molecular probing in laboratories, rather than the time and space consuming visual identification of the expressed trait. This allows plant breeders to screen large populations of plants for those that possess the traits of interest. However, the major contribution of Genetic Engineering has been in *Genetic Modification* of plants to produce *GM crops*. This is achieved by adding a specific gene or genes to a plant, or by gene silencing through knocking down of a gene with RNAi to produce a desirable GM crop. *The special feature of this technology is the transfer of genes across species, genera and even kingdoms.* This is achieved by splicing the gene DNA segment of a desired trait to a mobile carrier or vector DNA which can 'jump' across species. Such a Recombinant DNA construct enables the gene to cross the species barriers and integrate with the genome of the target plant causing its genetic modification. The plants resulting from such transformation are referred to as *Transgenic Plants*. The construct, basically the vector DNA with transgene, and other DNA segments, mainly a promoter for expression of the gene, markers for identification of transformed cells or tissues, can be inserted in the plant genome by genetic recombination using the bacterium *Agrobacterium tumefaciens,* or by direct methods like *particle bombardment* with gene gun. It may be mentioned at this point that the concept of genes 'jumping' from one chromosome to another was given long back in early 50s by Barbara McClintock. Working with corn, she had demonstrated the occurrence of some controlling elements which induced inter-chromosomal gene movements. At that time her concept was not accepted because of the rigid views about the 'fixed' nature of genes. However, the rediscovery of the controlling elements as *transposons* led to her recognition and award of Nobel Prize to her in 1985. The transposons also play important role in the transfer of genes across species.

The majority of commercially released transgenic plants are currently limited to plants that have introduced *resistance to insect pests and herbicides*. Genetic modification of plants that can produce *pharmaceuticals* and other industrial chemicals is also an area of active endeavour. If for genetic modification, genes of the same species or of a crossable plant are used under control of their native promoter, then they are termed as *cisgenic* plants. Crop improvement, whether classical or through genetic engineering, come with issues of concern, particularly with regard to food crops. The concerns encompass the *ecological impact* of GM crops, the *biosafety* of GM food and protocols used for safety evaluation like *substantial equivalence.* Most countries need stringent regulatory processes in place to help ensure that new GM crop varieties entering the marketplace are both safe and meet farmer's needs. There is active ongoing research to improve plant transformation technology with emphasis on gene stacking, site directed insertion of transgene, removal of negative markers, use of safe tissue specific promoters, signal sensitive promoters, prevention of gene 'jumping' and other aspects to make the GM product biosafe.

Plant Breeder's Right is also a major and controversial issue. The policy of allowing patenting of 'novel living organisms' has led to restrictions unlike the formerly available, open access to plant breeder's creations. An extension of the situation, specially in the case of GM crops is, the difficulty in incorporating technical advances in the production protocols. GM crops are grown, in the third world, by moderately rich and mainly small and marginal farmers. Price tags, due to patents, increases the production costs and the ultimate price of the product. In case of crops, specially food crops, special provisions, as government subsidies for patents, must be made for ensuring incorporation of technical innovations for safety, in the production protocols. This will make safe GM crop seeds affordable for our farmers and the farm products purchasable by consumers.

The agriculture improvement programme of our organization, Samaj Pragati Sahayog, spread over more than twelve states and nearly eighty districts of the country, has added new dimensions for progress in the farm sector. The main indication is that a holistic approach is needed. To obtain greater impact of improvements in crops and agriculture practices, farming has to be linked with watershed development interventions of soil and water conservation. Such interventions generate employment, thereby increasing the purchasing power of the rural masses. Further, linkage of farmers with banks and the market, through empowered self help groups of villagers, for procurement of goods and sale of farm products makes the farmers self reliant. We must remember, majority of our farmers, nearly 80%, are of small and marginal scale and 70 % of our country is dryland. Farming is mainly rainfed and most farmers can take only one crop. This is mainly because, much of the dryland, a part of the Deccan trap, extending from south to parts of MP in north, are underlain by basalt hard rock which is not amenable for bore wells. Therefore, as we have done, recharge of ground water through rain water harvesting, which mostly ensures some water availability in the wells and *in situ* soil moisture conservation even after rains, creates possibilities of a second crop. Thus, enhancement of productivity of the lands of all farmers, big, small and marginal, by the watershed development approach, is a practical way of achieving food security targets. There will have to be probably a shift in focus in crop improvement research, from high yields to abiotic stresses as drought resistance and biotic stresses as resistance to pests and diseases. With expanded gene resource and updated safe recombinant DNA technology, possibilities of developing a variety of stress resistant crops exist. A combination of watershed development approach for enhancing productivity of agriculture lands and farming with such crops, even with moderate yields, requiring low inputs, can usher in a true evergreen revolution.

3

Basic Aspects of Molecular Biology

SUMMARY OF CONTENTS

Cellular organization; Structure of basic macromolecules, DNA, RNA and Proteins; Prokaryotic and Eukaryotic gene; DNA replication, repair; Gene expression - transcription, RNA types, RNA processing, translation; Regulation of gene expression–prokaryotic and eukaryotic; Cell division in somatic and reproductive cells; Structure -function correlations of macromolecules.

Recombinant DNA technology involves manipulation of genes. Therefore, some familiarity with the general aspects of genes, gene structure, expression and regulation will be useful. In the following sections some of these aspects are described.

3.1. CELLS

Since long, DNA has been established as the chemical basis of heredity. The simplistic central theme is that DNA makes RNA, RNA makes protein and the proteins govern the traits (Fig. 3.1).

These processes happen in the cells of different organisms. The organisation of bacterial, plant and animal cells are shown in Fig. 3.2.

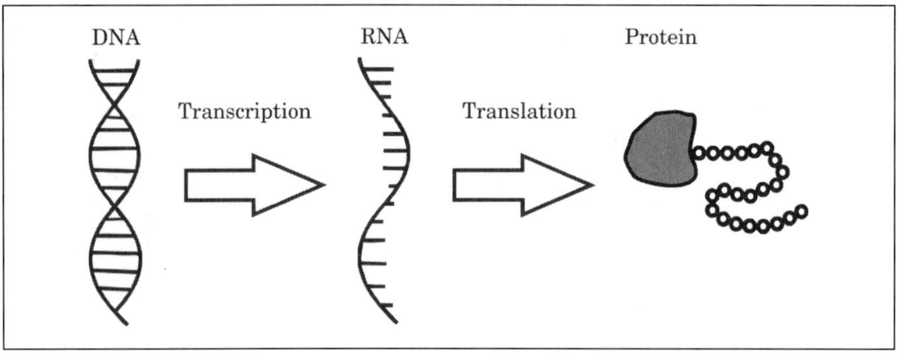

Fig. 3.1: Genes to traits, The central dogma

There are some major differences between cells of bacteria and those of the higher organisms as plant and animal cells. The DNA of a bacterium is circular, complexed to few proteins and is present as a nucleoid which is not enclosed in a membrane. Therefore, organisms as bacteria are grouped under Prokaryotes (Greek, *pro* = before, *karyon* = nucleus). Cells of higher organisms have several linear DNA strands complexed to many proteins and each such strand is organized into individual chromosomes. The chromosomes are enclosed within a membrane envelope constituting the nucleus. Organisms with nucleus are termed as Eukaryotes (*eu* = true, *karyon* = nucleus). Optical microscopic studies (Fig. 3.2) had revealed that besides the nucleus, all cells have other structures, the organelles. The plant cells mainly differ from animal cells in having the chloroplasts. The mitochondria are the 'power houses' of cells, carrying out respiration and releasing energy in the form of ATP (adenosine triphosphate) molecules. The chloroplasts are the 'kitchens' of plant cells, capturing solar energy and with the help of this energy synthesizing food stuffs as glucose. Electron microscopy revealed much more magnified view of the cells, the ultrastructure of cells (Fig. 3.2). Ultrastructural studies showed the occurrence of other organelles and their features. Thus, the cell consists of a network of microtubules, the endoplasmic reticulum; a mesh of fibres the cytoskeleton; organelles as Golgi body/dictyosome and ribosomes. Ribosomes are the seats of polypeptide synthesis. Lipid synthesis, some processing and transport of polypeptides take place in the

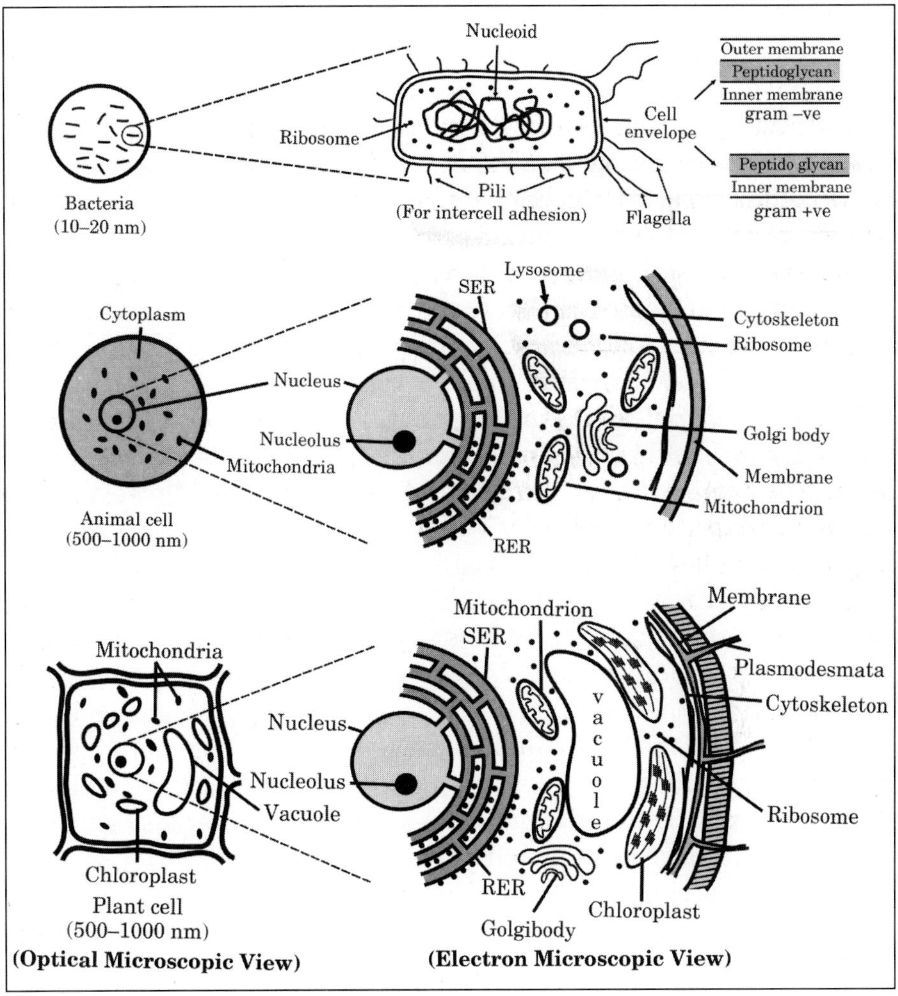

Fig. 3.2: Cellular organization of bacterial, plant and animal cells

endoplasmic reticulum. Golgi body helps mainly in modifications in proteins and their use and secretion. The cytoskeleton maintains cell shape and helps in movements of internal structures as the chromosomes during cell division.

The details of the cell structure have been revealed mainly through studies with optical and electron microscopes. However an important tool for isolation of individual organelles and macromolecules has been *differential centrifugation*. Centrifuge equipment, commonly available in most biolabs, consists of an angle rotor with cavities for keeping polyethylene or glass tubes symmetrically opposite each other in a circle, usually oriented at a fixed angle to the vertical axis of the rotor. Some rotors are of the swingout type, with flexible rings for keeping the tubes, as in a merry go round. The rotor is kept on the top of a motor. Spinning of the rotor subjects the contents of the tubes to a centrifugal force. In angle rotors the direction of the force is sideways out on the tubes and in swing out rotors straight outwards from top to bottom of the tubes. If a suspension is kept in a vertical tube, the particles of different sizes, pulled down by gravity, will sediment at different rates when passing *through a medium*. Larger particles will sediment faster than the smaller particles of same density. However, the sedimentation due to gravity will be at a very slow rate and the fragile organelles or molecules will be destroyed. In centrifuges the sedimentation rate is speeded up by applying a centrifugal force, many times that of the force of gravity. Therefore, the centrifugal force applied is indicated as, X g. To isolate organelles by differential centrifugation, the sample tissue is crushed or homogenised in a suitable liquid medium, equal volumes of the homogenate are put in centrifuge tubes, the tubes kept usually in an angle rotor and the rotor is spun. Different organelles sediment at different x g values, The order of sedimentation, with ordinary centrifuges (100 to 20000 x g range), is usually, nuclei → chloroplasts → mitochondria in plant cells. For instance, nuclei may sediment at 100 x g for 10 minutes. So, the homogenates is spun for 10 minutes at 100 x g. The nuclei sediment, settle on the outer sidewall near the bottom of each tube. The sediment or *pellet* is collected and the residual fluid the *supernatant* is subjected to further centrifugation for stepwise isolation of chloroplasts and mitochondria. For minute organelles as ribosomes and ER particles, the microsomes, ultracentrifugation (about 50000 x g upwards) is used.

For smaller particles and macromolecules and for separation of organelles of nearly similar sizes as separation of mitochondria from peroxisomes in a mixture, *density gradient ultra centrifugation*, in swing out rotors, is done. Examples of such centrifugation have been given in Chapter 4.

The building blocks of cellular structures are some macromolecules, the polysaccharides, nucleic acids, proteins and lipids. While the first three are macromolecules by virtue of being high molecular weight polymers, the lipids have high molecular weight but not a polymeric organization. Polymers are chains of monomers, that is similar low molecular weight compounds. Thus, starch, a polysaccharide, is a chain of glucose, nucleic acids are chains of nucleotides and proteins are chains of amino acids. Lipids are a diverse group of chemicals, ranging from cholesterol to triglycerides. In the following sections the main macromolecules involved in recombinant DNA technology, nucleic acids and proteins have been described.

3.2. NUCLEIC ACIDS (DNA & RNA)

DNA (Deoxyribonucleic acid) occurs mainly in the nucleus and also in the chloroplasts and mitochondria. While DNA is the storehouse of genetic information, RNA (Ribonucleic acid) helps in transfer and translation of this information into polypeptide sequences.

The main DNA resides in the nucleus. Complementary RNA molecules, as messenger RNA (mRNA), ribosomal RNA (rRNA) and transfer RNA (tRNA) are synthesized on DNA templates in the nucleus and move out to the cytoplasm. In the cytoplasm, the proteins are synthesized on mRNA molecules with the help of rRNA containing ribosomes and tRNA. It is known that each protein chain has a specific amino acid sequence. Synthesis of proteins of such specific sequences is dependent and ensured by the genes in DNA molecules. Thus, each gene is a blueprint for a specific protein sequence. Gene expression is the encoding of a protein/s by a gene under the control of various regulatory elements.

Nucleic acids are polynucleotides, each molecule being a polymeric chain of nucleotide monomers. A nucleotide consists of a pentose sugar moiety to which is linked a phosphate residue and a base. There are four bases, Adenine (A), Thymine (T), Guanine (G) and Cytosine (C) in DNA and A, G, C, and Uracil (U) in RNA. RNA is a chain of ribonucleotides and DNA is a chain of deoxyribonucleotides. The nucleotides are linked by phosphodiester bonds. Main features of the primary structures of DNA and RNA are shown in Fig. 3.3.

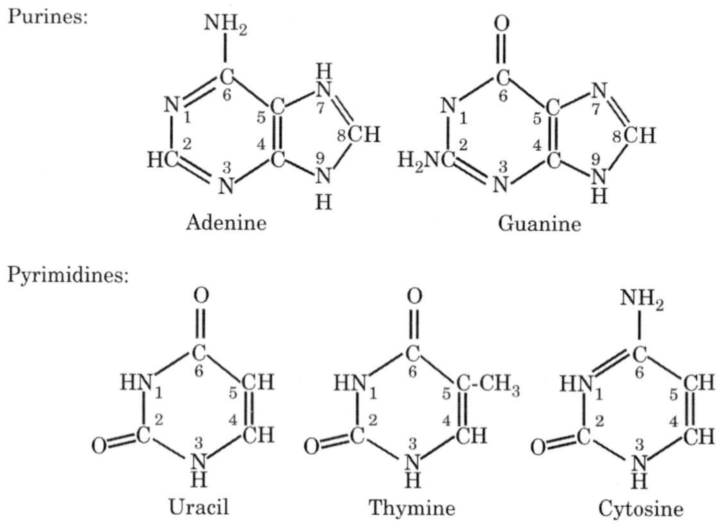

(a) Nitrogenous bases

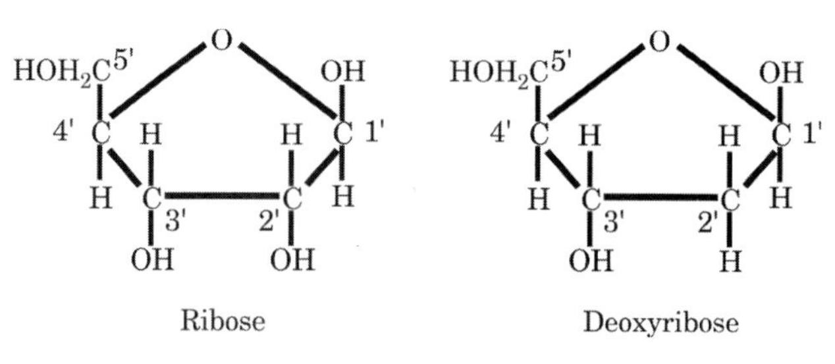

(b) Pentose sugars

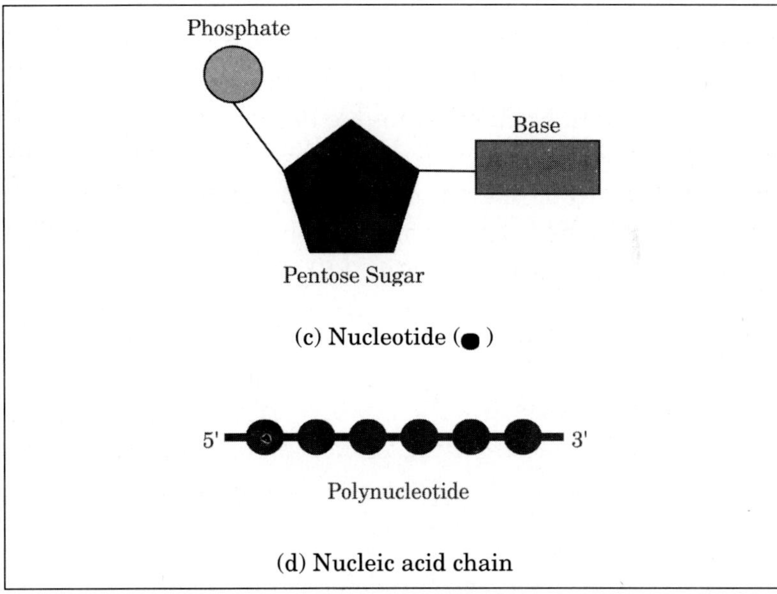

Fig. 3.3: Features of the primary structures of DNA and RNA

The differences between DNA and RNA are shown also in Fig. 3.4 below.

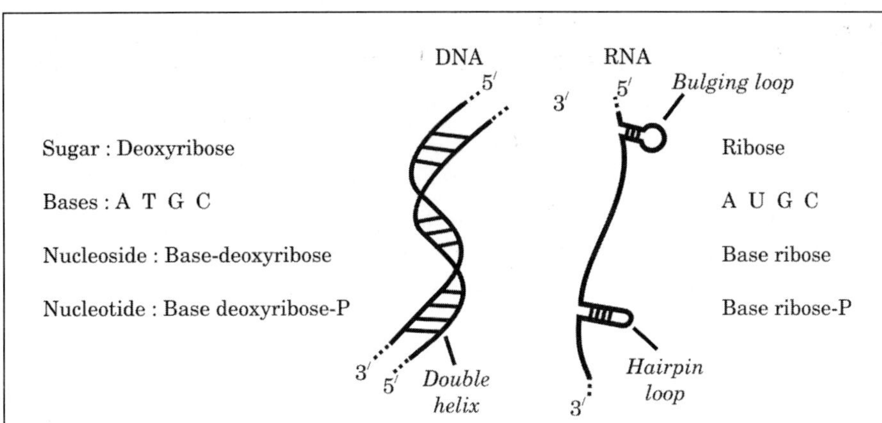

Fig. 3.4: Differences between DNA and RNA

Deoxyribonucleic Acid

A DNA molecule has two antiparallel strands which twist in a double helical conformation. Each such double helix has a sugar phosphate backbone and a core of hydrogen bonded base pairs. The two strands are held together by H-bonds between complementary bases, A-T and G-C. Such H bonded strands of complementary bases are referred to as complementary strands (Fig. 3.5).

The synthesis of a new DNA chain or *replication*, occurs on a prexisting DNA strand acting as a template for assembly and bonding of complementary bases from a mixture of A, T, G, C deoxyribonucleotides. A DNA chain thus, makes a complementary strand. Similarly, a DNA chain, acts also as a template for the synthesis or *transcription* of a complementary RNA strand.

In the cells, DNA exists as a nucleoprotein complex. It is mainly complexed with basic proteins as histones and some acidic proteins to form chromatin fibres. Each chromosome is a long DNA molecule complexed to many proteins. A chromosome essentially is a condensed nucleoprotein chromatin fibre. Actually, in a resting cell, the chromosome chromatin fibres are microscopically indistinguishable from each other. Prior to cell division the chromatin fibres condense and each chromosome becomes visible. Each fibre in the extended form, is a chain of nucleosomes. Each nucleosome is a core complex of mainly eight histones, two each of H2A, H2B, H3 and H4, wrapped around by DNA of about 150 bp length. The nucleosomes are separated by spacer DNA with H1 histone attached to it. As a cell approaches the stage of mitotic or meiotic division, the chromatin condenses by stacking of the nucleosomes (Fig. 3.6).

It must be noted that in the DNA → RNA → Protein flow chain, a DNA strand acts as a template for the mRNA and the mRNA, a template for the protein chain.

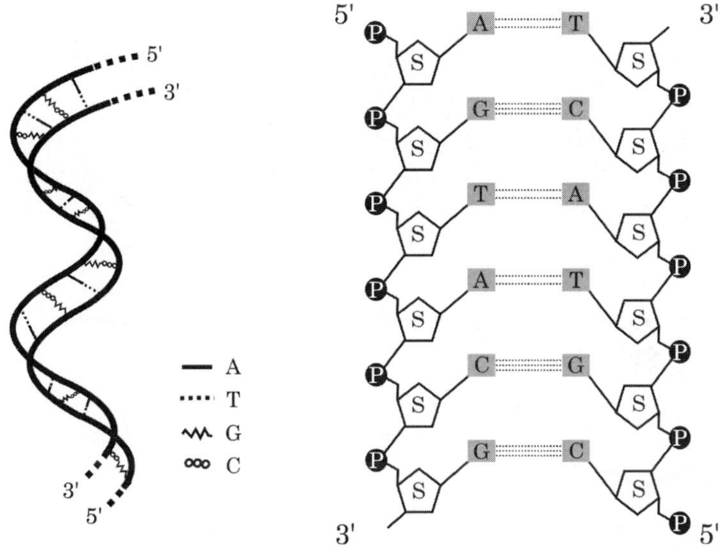

a. DNA double helix **b.** DNA antiparallel strands

Fig. 3.5a & b: The secondary structure of DNA

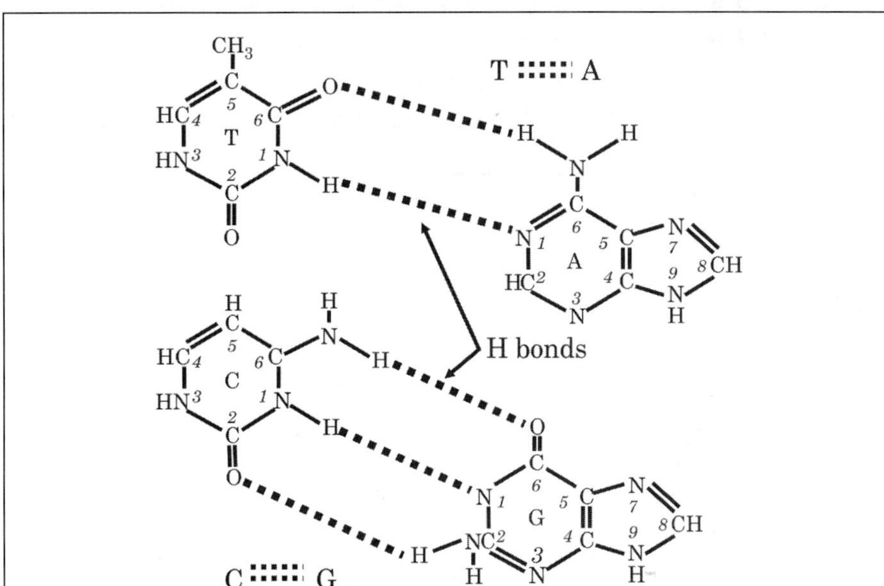

Fig. 3.5c: Hydrogen bonds between purines and pyrimidines

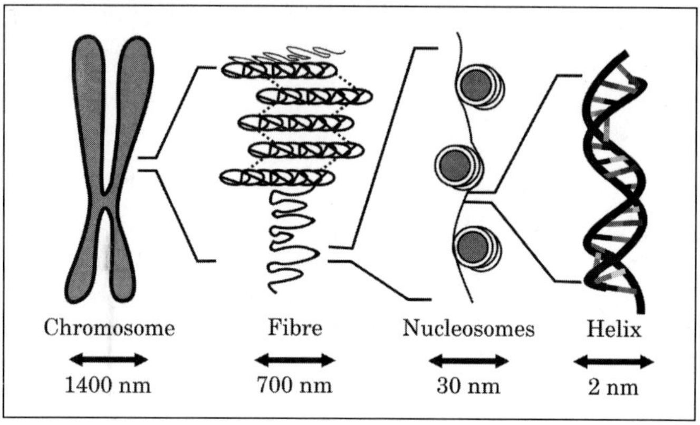

Fig. 3.6: Chromatin structure

Ribonucleic Acid

RNA, like DNA, is a polymer of the nucleotides A, G, C and uracil (U) instead of thymine (T). There are three main types of RNA. The messenger RNA (mRNA), ribosomal RNA (rRNA) and transfer RNA (tRNA). Messenger RNA is a sequence of triplet codes, codons, for specific amino acids (Table 3.1). Ribosomal RNA forms the backbone of the ribosomes. Ribosomes form the platform on which protein synthesis takes place. Messenger RNA attaches to ribosomes to form polysomes the fundamental units for protein synthesis (Fig. 3.7).

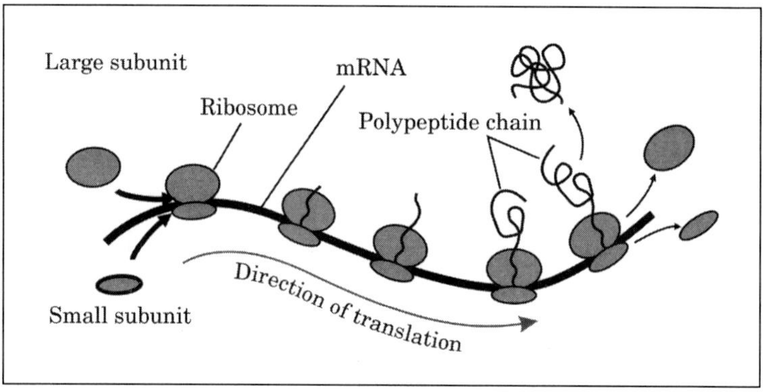

Fig. 3.7: Polysomes showing mRNA and ribosomes

The term transfer RNA relates to the function of these RNAs. Each transfer RNA carries a specific amino acid to the polysome, the protein synthesis machinery (Fig. 3.8).

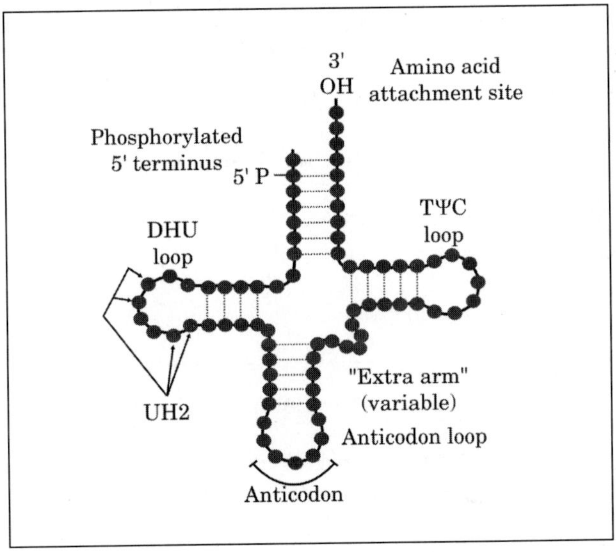

Fig. 3.8: Transfer RNA

The different types of RNA and their functions are described in greater detail in the section on translation below.

3.3. PROTEIN

A cell contains numerous proteins with different functions. Proteins constitute many functional and structural molecules. Functional proteins include the enzymes, hormones, transcription factors and others. Examples of structural proteins are tubulin and actin which make up the cytoskeleton.

A protein molecule consists of a single or several polypeptide chains. Each protein chain is a polymer of amino acids, which are linked by peptide bonds and hence the term polypeptide (Fig. 3.9).

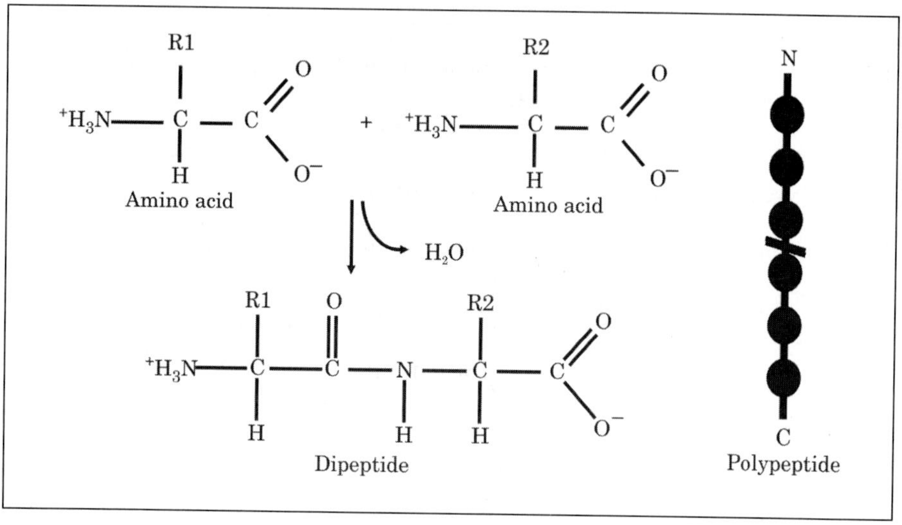

Fig. 3.9: Protein structure: Amino acids and peptides

Polypeptides exist in several structural forms as secondary, tertiary and quaternary forms (Fig. 3.10).

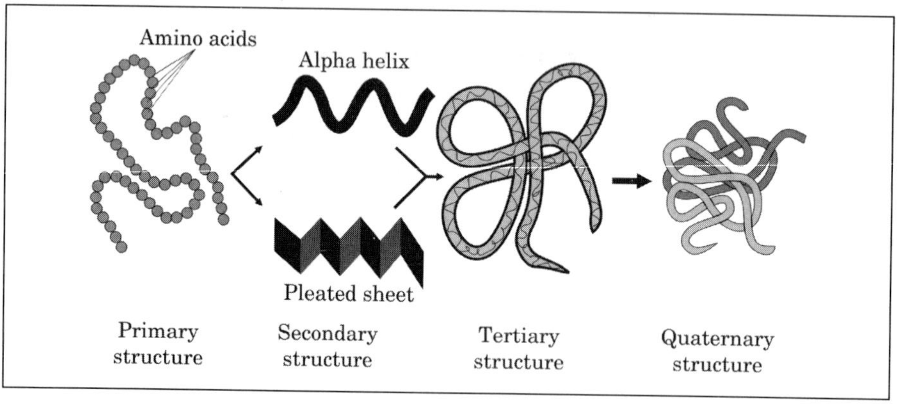

Fig. 3.10: Structural levels of proteins

Alpha heilcal and β-pleated sheet are common secondary structures (Fig. 3.11).

A specific folded tertiary conformation is necessary for optimal functionality of proteins as, enzymes. Such conformations are

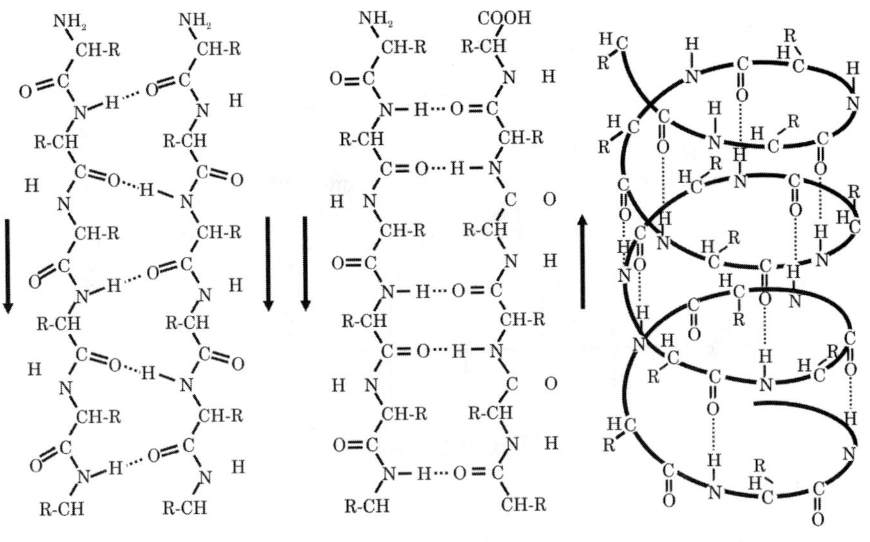

Fig. 3.11: Secondary structural forms of proteins

maintained by several forces as H bonds, ionic forces of attraction, Van der Waal's forces and disulphide bonds (Fig. 3.12).

There are 20 amino acids each with specific trinucleotide codes, the triplet codes in the DNA coding strand (Table 3.1).

In the above table amino acids have been shown with traditional three letter abbreviations. Now, however, single letter abbreviations are used as: Alanine (A), Arginine (R), Asparagine (N), Aspartic Acid (D), Cysteine (C), Glutamic Acid (E), Glutamine (Q), Glycine (G), Histidine (H), Isoleucine (I), Leucine (L), Lysine (K), Methionine (M), Phenylalanine (F), Proline (P), Serine (S), Threonine (T), Tryptophan (W), Tyrosine (Y), Valine (V).

Every protein chain has a specific amino acid sequence. During protein synthesis or *translation* each amino acid is recognized by a sequence of three nucleotides, the nucloetide triplet code or codon,

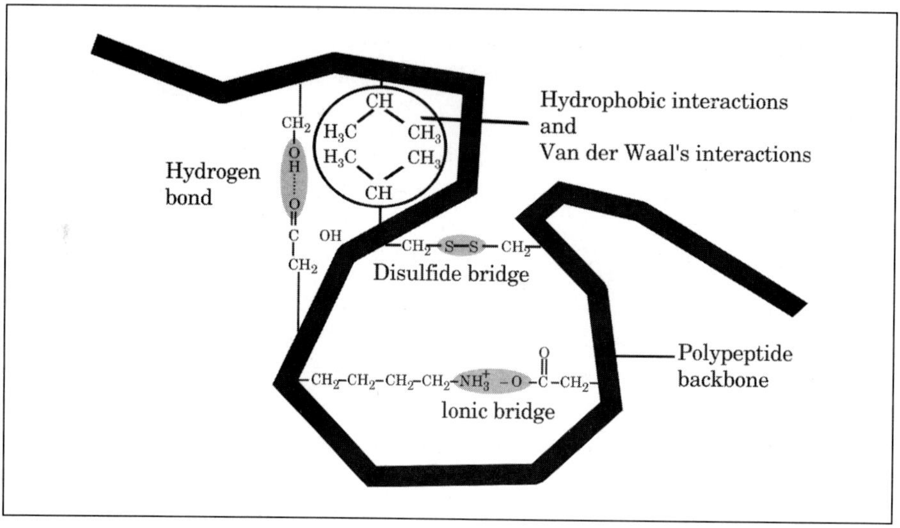

Fig. 3.12: Forces stabilizing the tertiary structure of proteins

Table 3.1: Triplet codes for protein amino acids

Triplet codes for protein amino acids							
UUU	Phe	UCU	Ser	UAU	Tyr	UGU	Cys
UUC	Phe	UCC	Ser	UAC	Tyr	UGG	Cys
UUA	Leu	UCA	Ser	UAA	Stop	UGA	Stop
UUG	Leu	UCG	Ser	UAG	Stop	UGG	Trp
CUU	Leu	CCU	Pro	GAU	His	CGU	Arg
CUC	Leu	CCC	Pro	CAC	His	CGC	Arg
CUA	Leu	CCA	Pro	CAA	Gln	CGA	Arg
CUG	Leu	CCG	Pro	CAG	Gln	CGG	Arg
AUU	Ile	ACU	Thr	AAU	Asn	AGU	Ser
AUC	Ile	ACC	Thr	AAC	Asn	AGC	Ser
AUA	Ile	ACA	Thr	AAA	Lys	AGA	Arg
AUG	Met*	ACG	Thr	AAG	Lys	AGG	Arg
GUU	Val	GCU	Ala	GAU	Asp	GGU	Gly
GUC	Val	GCC	Ala	GAG	Asp	GGC	Gly
GUA	Val	GCA	Ala	GAA	Glu	GGA	Gly
GUG	Val	GCG	Ala	GAG	Glu	GGG	Gly

occurring in the mRNA transcribed from the template strand. Thus, if AAA is the codon for lysine (Lys) and GGG for Glycine (Gly), for synthesis of the protein chain N....Gly.Gly.Lys.Lys....C, an mRNA template strand of the sequence 5′....GGG.GGG.AAA.AAA....3′ will be required. And, this mRNA will be synthesized on a DNA template strand of sequence 3′...CCC.CCC.TTT.TTT....5′. The other DNA strand which has the actual sequence of deoxyribonucleotide triplet code sequence in the 5′→ 3′ direction is the coding strand of the DNA duplex. *A gene sequence is the sequence of the coding DNA strand.* It must be noted that this coding strand does not act as the template strand for mRNA transcription. The template DNA strand has the complementary sequence and leads to the synthesis of mRNA which has the correct codon sequence or blueprint of the amino acid sequence of the polypeptide to be synthesized, like the coding DNA strand (Fig. 3.13).

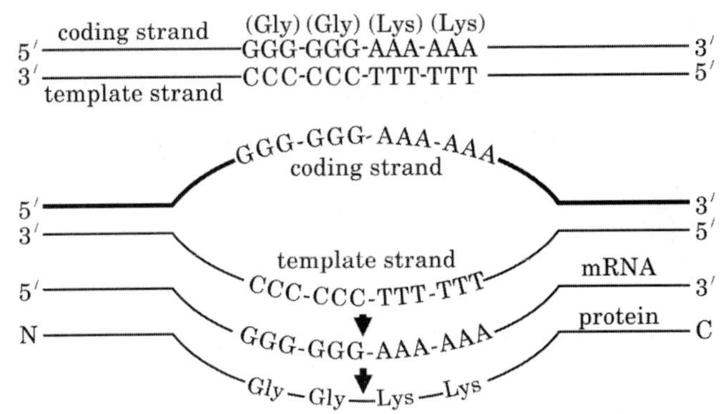

Fig. 3.13: DNA template and coding strands

3.4. GENE

The above description gives an idea of a gene. A gene in its simplest form is a sequence of DNA codons which encode a specific protein chain and some regulatory sequences specific to it. Exceptions are the genes for ribosomal and transfer RNAs. The protein coding genes can be of two types, the structural genes and regulatory genes. The products of structural genes include proteins which are components of structures and also proteins which perform some specific functions as enzymes and hormones. The regulatory genes are involved in the regulation of gene expression. The length of genes can range from hundreds to thousands of nucleotides. As has been shown in Table 3.1 the triplet codes for amino acids, the *genetic code* is degenerate, that is generally, there are more than one codon for each amino acid.

Genes have evolved over time and the evolution of genes underlie evolution of organisms. The prokaryotes have the simplest usually uninterrupted codon sequences as genes, while the eukaryotes have more complex genes with the codon sequence interrupted.

Prokaryotic Gene

A prokaryotic gene in its simplest form consists of a codon sequence with regulatory elements as promoter and operator sequences on the 5′ side. The promoter sequence is like a switch for the coding region and the operator is involved in switching the promoter on or off.

Such genes which encode one protein are termed as monocistronic, the coding region being the cistron. There are also polycistronic genes, in which the promoter and operator are followed by several coding regions or cistrons in a row adjacent to each other. These features are shown in Fig. 3.14.

Eukaryotic Gene

The advent of recombinant DNA technology, functional genomics and genome sequencing studies have revealed the basic features of eukaryotic genes including plant genes. The essential features are shown in Fig 3.15.

Typically a protein coding gene consists of a sequence of codons which determine the amino acid sequence of a protein chain via a complementary mRNA. As mentioned earlier, by transcribing on the

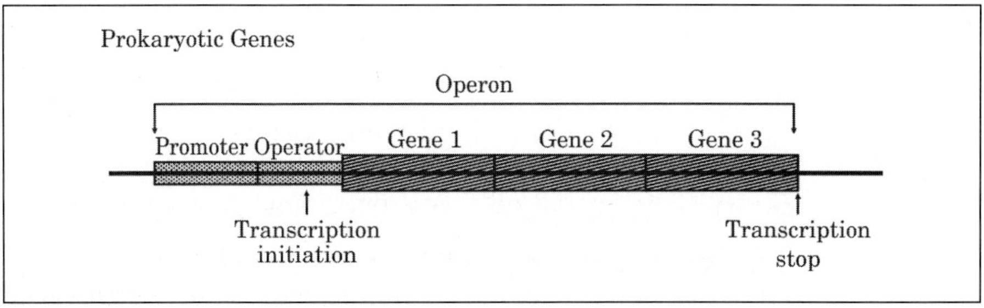

Fig. 3.14: The prokaryotic gene

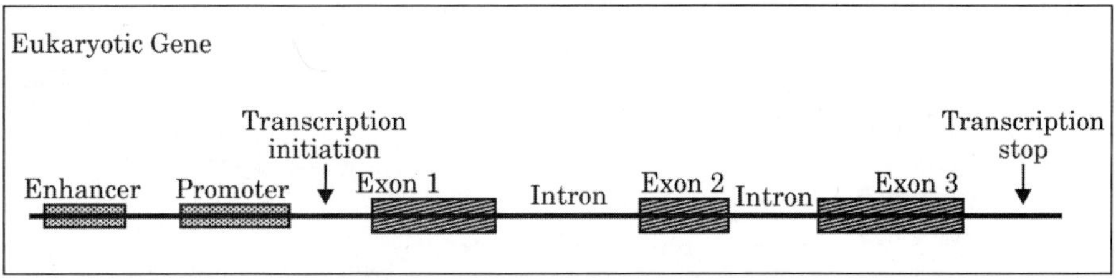

Fig. 3.15: The eukaryotic gene

template DNA strand, the needed codon sequence is transferred to the mRNA formed. The gene consists of a coding sequence with the transcription regulatory sequences flanking the 5′ side and termination and polyadenylation sequences on the 3′ side (with the coding strand as the reference). The starting point of transcription, the *start site* is numbered +1. Towards the 5′ side of this start site is the *promoter*. The promoter region consists of a TATA sequence, the TATA box, 20 to 25 bases from the start site and a CAAT box further 70 to 80 bases up. In plants the TATA sequence is TCACTATATATAG and for the CAAT set, there is an alternate AGGA box also. Between start site and the beginning of the coding region, and also the end of the coding region and the termination site are stretches of about 60 to 80 bases which are not codons. These stretches before and after the coding regions give rise to 5′ untranslated and 3′ untranslated regions (5′UTR and 3′UTR regions) in the mRNA. The coding region has codon containing sequences the *exons*, interrupted by noncoding sequences, the *introns.*

3.5. DNA REPLICATION

Growth of organisms whether prokaryotes or eukaryotes is dependent on cell divisions. Prior to the cell division process, the cellular components have to be increased and duplicated to be passed on equally between the two daughter cells. This is specially so for the chromosomes and their constituent DNA. The process of DNA duplication is termed as *replication.* In this process each strand acts as a template for formation of a complementary strand. The replication is therefore, *semiconservative*, each double stranded product containing a parent and a new strand (Fig. 3.16).

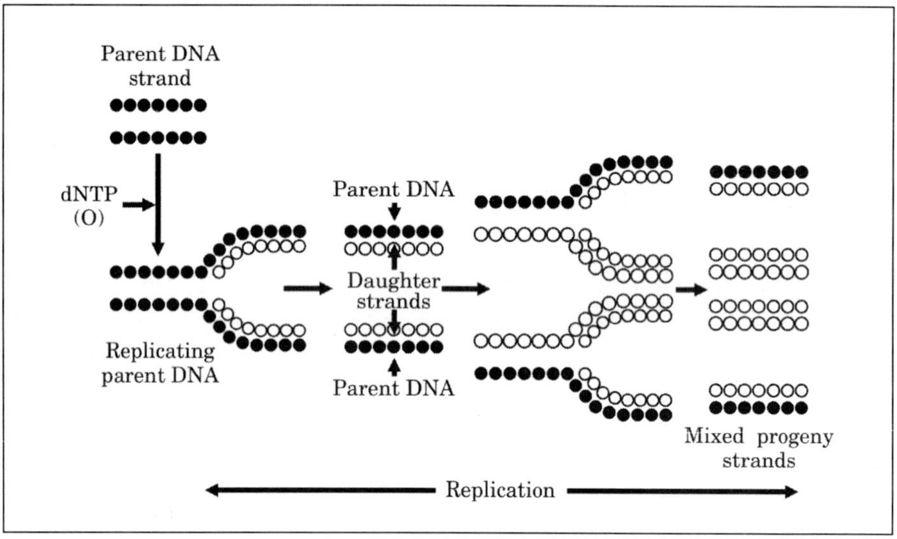

Fig. 3.16: Semi conservative DNA replication

Generalized steps of replication are shown in Fig. 3.17.

The replication starts at a site of about 150 bases, referred to as the *origin of replication* (ori). The synthesis of a chain is in the $5'$ to $3'$ direction. Enzyme and multiprotein complexes containing units as helicase and gyrase or *DNA topoisomerase* attach to the DNA in the vicinity of ori and unwind the helix in a stabilized *replication fork* conformation. While helicase unwinds the helix, gyrase stabilizes the replication fork by preventing supercoiling. The

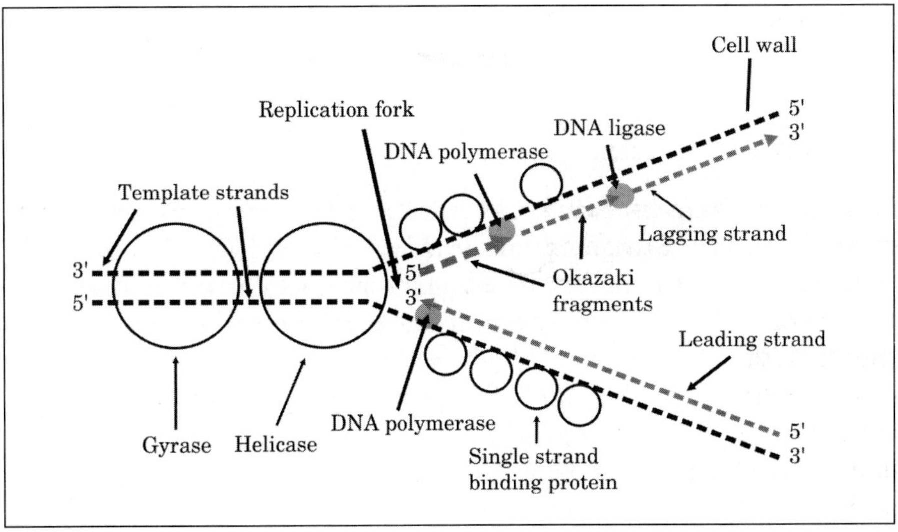

Fig. 3.17: The replication process

unwinded strands are prevented from reannealing by some SSB proteins (single strand binding) attaching to each strand. These proteins get removed as replication proceeds. On both strands near ori, short RNA oligonucleotides, the *primers* are synthesized by *primase* enzyme. Now DNA polymerase (different types of DNA polymerase in bacteria and eukaryotes) keeps attaching deoxyribonucleotides to the $3'$OH ends of the primers, in a linear array. On one of the strands, the site of attachment of deoxynucleotides, that is the $3'$OH end, is continuously available due to the opening of helix. Hence, it is called the *leading* strand, where the synthesis is continuous. In the other *lagging* strand the synthesis is in short stretches, due to opening of DNA helix towards the $5'$ end, where, new deoxynucleotides cannot be added. This leads to the formation of short oligonucleotides called Okazaki fragments. Synthesis in short stretches takes place because on the $5'$ side, nucleotides cannot be added. Ultimately these strands are joined together by DNA ligase enzyme to form a continuous strand. The replication ends at termination sequence points. It may be noted that bacteria usually have one ori for a gene while an eukaryote gene has many. Another interesting feature is that the DNA polymerase does twin task of *synthesis* as also *proof reading* through its exonuclease activity.

DNA Repair

DNA in the cells is subject to damages by various internal and external factors. Thus, ionizing radiations, UV radiations, certain chemicals and other agents can damage the cellular DNA. Even heat produced during metabolism can remove adenine and guanine from the DNA. Such damages or *lesions* in somatic cells can lead to adverse consequences as malignancy. Lesions in the germ cells can pass on to the progeny. Nature, however, has mechanisms for repairing such damages to the DNA backbone. There are different types of damages and related repair mechanisms. Some major types are briefly described here.

A lesion can occur in the nucleotides of a gene. Such a damage is repaired by the *nucleotide excision repair* system. A transcription factor TFIIH, an oligomeric protein, recognizes the damage, an endonuclease cuts and removes the altered nucleotide/s, DNA polymerase incorporates the correct nucleotide/s and a ligase joins the broken strands.

Damage in a base is repaired by the *base excision repair* system. The damaged base is recognized and removed by a specific DNA glycosylase. This is followed by the removal of the residual deoxyribosephosphate by a combined action of an AP endonuclease and the nuclease fragment of β DNA polymerase. Thereafter, the proper nucleotide is filled in by β DNA polymerase and the strand joined by ligase.

Distortion in the DNA strand can occur due to mismatch between complementary base pairs. Therefore, mechanisms of *mismatch repair* also exist. Although, in higher organisms the exact system is under investigation, it has been found that in *E. coli*, this mismatch is due to the fact that while the adenosine moiety in the parental strand is methylated, in the daughter strand it is not methylated.

Some Mut proteins recognize the mismatch, a helicase unwinds the mismatched section, an exonuclease removes it and the gap is filled by DNA polymerase I and ligase.

Double strand breaks constitute a major lesion in the DNA backbone. These are caused by various radiations, chemicals and products of metabolism as free radicals. The main mechanism of repair consists of *nonhomologous end joining*. In some higher organisms, the repair starts with recognition of the break by a heterodimeric protein Ku, which binds to the break region. A DNA-PK_{cs} protein kinase fragment, brings the broken ends in proximity and a ligase joins them.

3.6. GENE EXPRESSION

Expression of a gene, a segment of DNA, occurs via the processes of complementary messenger RNA synthesis, *transcription* and synthesis of a protein chain on this mRNA template with the help of ribosomal RNA (rRNA) and transfer RNA (tRNA), the process of *translation*.

Transcription

Transcription is the process of mRNA synthesis on a DNA template strand. Its codon sequence is a complement of the template strand and an exact copy of the coding strand of the gene (with U replacing T) for translation into its protein product. The process requires the promoter sequences on the template strand and some proteinaceous general transcription factors, GTFs, and specific transfer factors, STFs. Besides these, some DNA sequences, the enhancer elements, about even 1kb upstream of the *start site* are also needed. The regulatory sequences as the promoter and enhancer elements of the DNA strand are referred to as *cis regulatory elements* as they occur in the same strand and act on the coding region from nearby sequences or come close by folding of the chain. The GTF and STF are proteins present in the nuclear matrix and are not components of the DNA strand and therefore are regarded as the *trans regulatory elements*. The RNA polymerase II enzyme along with GTFs attach with the TATA box to form the *preinitiation complex*. The H bonds between the DNA strands at this point start breaking and the strands start to separate. The template surface gets exposed and becomes accessible for the attachment of the ribonucleotides to the complementary DNA bases and their progressive joining by RNA polymerase II, to form the mRNA. The process is further fine tuned by the CAAT box, the enhancer elements and the STFs. The protein transcription factors act by complexing with both the DNA strand and the RNA polymerase enzyme or other transcription factors. For this, each transcription factor has two sites or domains, a DNA *cis* domain for attaching to DNA and a *trans* domain for complexing with the enzyme or TF protein. For such specific associations the domains have specific conformations or motifs. Thus, the cis DNA domain has several motifs as the zinc-finger or leucine-zipper motif. The product of transcription, the transcript, is termed as *pre-mRNA*. It has from 5′ to 3′ end mainly, the 5′UTR region, the coding region of exons and introns, and the 3′ UTR region at the 3′ end. This *pre-mRNA* or *heterogeneous nuclear RNA* (*hnRNA*) has to undergo processing to form the mRNA needed, as template for *translation* of a protein chain.

The above account shows that binding of DNA with the polymerase and TFs lead to separation of the strands making the template strand accessible for transcription. In fact, the starting point of transcription is chromatin remodeling, uncoiling of DNA complexed with histones in the nucleosome. For this, some multiprotein units interact with nucleosomes and through energy from ATP breakdown, shift the nuclosomes to new positions to expose the DNA wrapped around the histone complex. Further, the enzyme *histone acetyl transferase* (HAT) acetylates the histones. Acetylated histones get removed from the DNA leading to uncoiling and accessibility of the template strand. On completion of transcription, *histone deacetylase complexes* (HDACs) deacetylate the histones leading to histone DNA complexing again and restoration of the nucleosome format of chromatin. The steps of transcription are shown in Fig. 3.18.

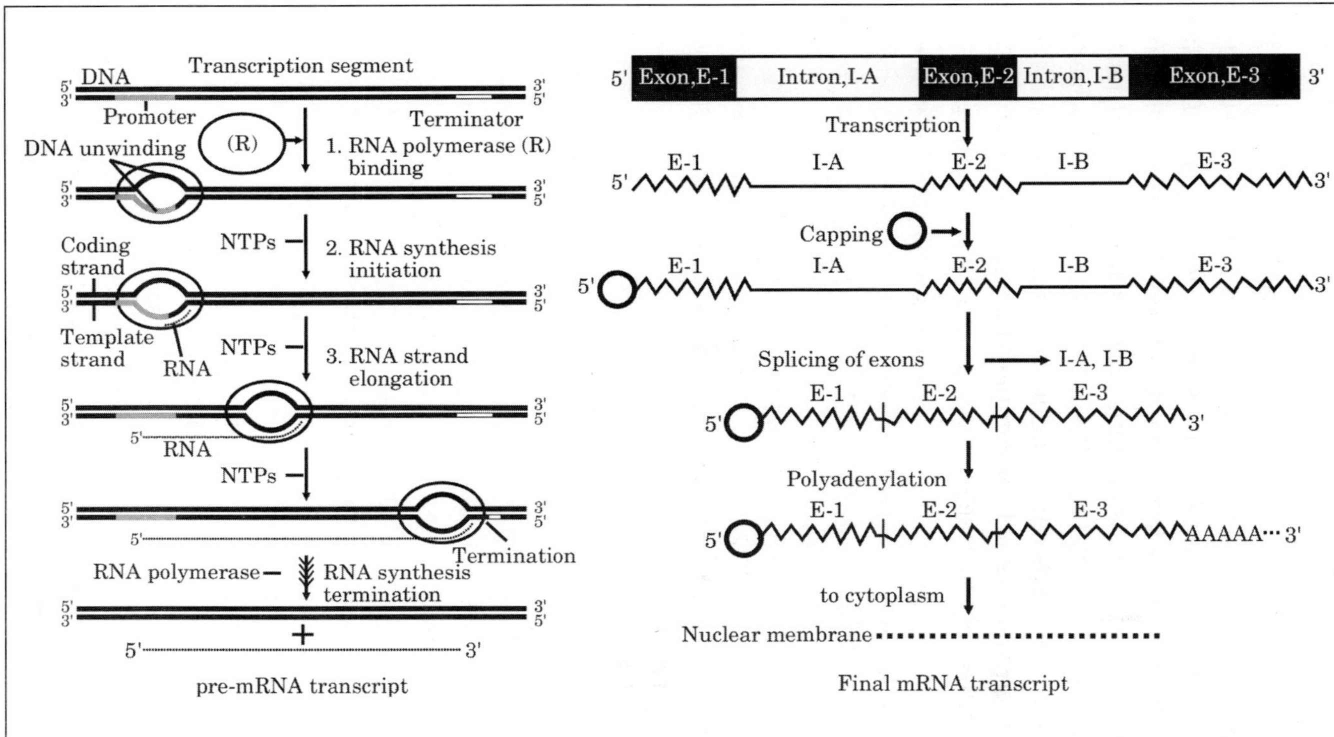

Fig. 3.18: The transcription process

Processing of pre-mRNA

Processing involves addition of a $5'$ cap and a $3'$ polyA tail, removal of *introns* and joining or splicing of *exons* (Fig. 3.18). At the $5'$ end, a 7-methyl guanosine is added. This *cap* acts as a protection against $5' \rightarrow 3'$ exonucleases. Similarly, about 200 bases long polyA tail is added at the $3'$ end. This may provide protection against $3' \rightarrow 5'$ exonucleases. Intron removal and exon splicing happen at the consensus sequences occurring in both the $5'$ exon-intron junction and the $3'$ intron-exon junction. The former are usually AG/GURAGU, and the latter YAG/RNNN (/ marks splicing site, R a purine, Y a pyrimidine, N any base). The process is assisted by spliceosomes which are complexes of small nuclear RNA (snRNA, about 100–300 bases long). The processed mRNA moves out of the nucleus for carrying out translation.

The splicing reactions may vary leading to shuffling of exons in different orders. Through such *alternate splicing* a gene may give rise to different proteins.

Synthesis and processing of other types of RNA

Ribosomal RNA (rRNA) exists in large amounts in the cell, to the extent of being 80% of total cellular RNA. It is transcribed by clusters of repeated units of rDNA. The initial transcript, pre-rRNA, is 45S size (about 13000 bases long) and is processed to yield 28 S (about 5000 bases), 18S (about 2000 bases) and 5.8S (about 160 bases) RNAs. This processing, in the interphase, occurs in the nucleolus with the help of small nucleolar RNAs (snoRNAs) which are complexed with proteins to form small nucleolar ribonucleoproteins (snoRNPs). Another RNA, the 5S RNA (about 120 nucleotides) is synthesized outside the nucleus and the transcript enters the nucleolus for ribosome assembly. The transfer RNA (tRNA) is encoded by clusters of repeated genes and the final tRNAs are derived through processing of a pre-tRNA through cleavage by some endonucleases and specially a RNaseP a part of which is itself RNA. In eukaryotic cells, in the nucleolus, the 60S and 40S subunits of ribosomes are assembled. The 60S unit contains about 50 proteins and 28S,5.8S and 5S RNAs and the 40S subunit has around 30 proteins and the 18S RNA.

It is worth noting that there are different RNA polymerases for the synthesis of different types RNA. RNA polymerase I transcribes, 28S, 18S and 5.8S rRNAs, RNA polymerase II synthesizes mainly mRNAs, snRNAs, snoRNAs and microRNA, RNA polymerase III brings about synthesis of tRNAs, 5S RNA and U6 RNA and in plants, poymerase IV transcribes siRNAs.

Ribozymes and RNA splicing

Thomas Cech in 1981, for the first time, demonstrated catalytic activity of RNA molecules in the splicing reaction in *Tetrahymena*. In this organism, a pre-mRNA spliced itself. This was an example of group I intron. Soon a group II type of self splicing intron was discovered in certain microbes, fungal mitochondria and chloroplasts. These RNA strands could also carry out self splicing. In eukaryotes, instead of self splicing, certain small nuclear RNAs, snRNAs are involved in the splicing of pre-mRNA. A complex of snRNAs and proteins form the spliceosome which attaches to introns and brings about splicing. The snRNAs are of different types and belong to the U series of snRNAs. Cech in 1981, on the basis of catalytic properties had termed such an RNA as *Ribozyme*. Ribozymes have been shown to have mainly splicing and peptidyl transferase activity.

The discovery of RNA enzymes, the ribozymes, changed the views about the chemical form in which life had originated. Previously, proteins and DNA were the candidates. This was because, only proteins were known to have enzymic activity and DNA possessed the codon information. However, the RNA ribozyme molecules have both, catalytic activity and codon information. Therefore, the present emphasis is on an early life of ‘RNA world’.

Small interfering RNAs (siRNA) and microRNA (miRNA) in gene silencing

Possibilities of the occurrence of small interfering RNAs was indicated in 1990s, when attempts to increase pink pigmentation in *Petunia* by addition of genes led to the appearance of white flowers rather than deeper pink flowers. Fire and Mello, in late nineties, clearly demonstrated the occurrence of small, about 20 nucleotide long, ds RNAs which silenced gene expression by hybridizing with the mRNA of similar sequence. This was mainly because of complexing of an mRNA in the proper orientation or 'sense' orientation with the 'antisense siRNA strand. Now, the occurrence of siRNAs is well established. The siRNAs arise under the influence of alien DNA, like those of viruses or transposons. These RNAs therefore, seem to be involved in the protection of an organism against alien nucleic acids, like viruses and transposons. In recent times, another class of small RNAs, again about 20 nucleotide long, the microRNAs (miRNAs) have been discovered in many organisms. These are encoded by the normal genome of an organism and seem to play a major role in the regulation of cellular metabolism.

Translation

The crucial point about the synthesis of a protein chain is ensuring that the linear sequence of amino acids from the N-terminal to the C-terminal end in the product is correct. Even change of one amino acid in the chain can lead to a change in its property. For such sequence specificities nature probably evolved the method of synthesis of copies of a template strand based on affinities, based on physical forces of attraction as H bonds, between chemical components of the strand and similar chemicals present in the ambient medium. Thus, synthesis of DNA to DNA or DNA to RNA (even RNA to DNA) are based on the affinities between A and T (or U) as also between G and C. In the cells, these bases collect on the template in an array, based on their affinity and are linked to form a copy by the requisite machinery. However, in the case of protein, a polypeptide the template is mRNA, a polynucleotide. So, where is the question of affinity or connection between nucleotides and amino acids? It has been mentioned that there are triplet codes, codons which recognize amino acids. Various combinations of the 4 bases have led to the formation of 3 nucleotide base based triplet codons. Thus, for 20 protein amino acids there are $4^3 = 64$ codons, each amino acid having several codons (Table 3.1). But, how the amino acids and their codons recognize each other or do they? In all organisms, the proper alignment of amino acids during protein synthesis is achieved by the combined action of transfer RNA (tRNA), ribosomes and the mRNA. Transfer RNA is synthesized by tRNA genes and is responsible for *carrying an amino acid to its proper codon*. Ribosomes consist of a RNA core complexed to proteins. There are two ribosomal units. The big ribosomal unit contains 28S rRNA, 5.8S rRNA and 5S rRNA, besides 49 proteins and the small ribosomal unit has 18S rRNA and 33 proteins. The translation process is shown in Fig. 3.19. The tRNAs are *amino acid specific*, have a clover leaf structure and an *anticodon sequence* in one of the loops (Fig. 3.7). The anticodon sequence is a complement of a codon. Thus, for a triplet codon ACC the anticodon will be UGG. An amino acid is attached by acyl bond formation between its –COOH and the 3′OH group of the tRNA terminal arm catalysed by aminoacyl-tRNA synthase. A tRNA so attached to an amino acid is termed as 'charged'. *Therefore, the recognition of a codon by an amino acid is indirect and is through the anticodon of the tRNA to which it gets attached*. The attachment to the specific tRNA is governed by the aminoacyl-tRNA synthase. The ribosomes bind to the mRNA and travel from 5′ end to the 3′ end while carrying out polypeptide synthesis. The ribosome units bind to the AUG initiation site. The large ribosome unit has three sites A (amino acyl), P (peptidyl), and E (exit) sites. The first tRNA charged with methionine, and with the anticodon UAC for AUG initiation codon, attaches to the first A site. The ribosome moves to the next codon. Now, as the tRNA charged with the next amino acid A1 comes, the met-tRNA shifts to the P site and a peptide bond is formed between met and A1. The released first tRNA now shifts to the E site, exits and the ribosome moves to the next codon. Again,

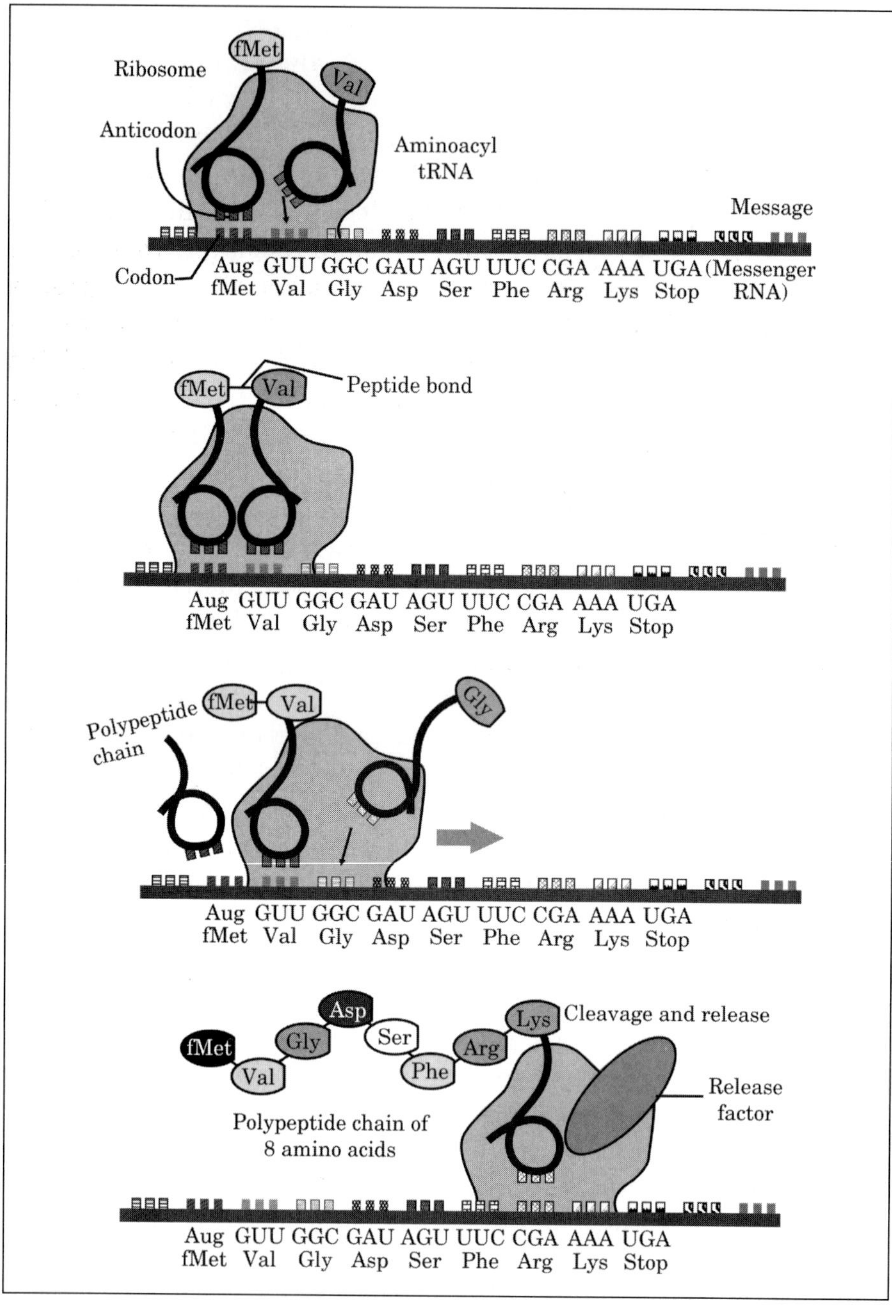

(a) Translation steps

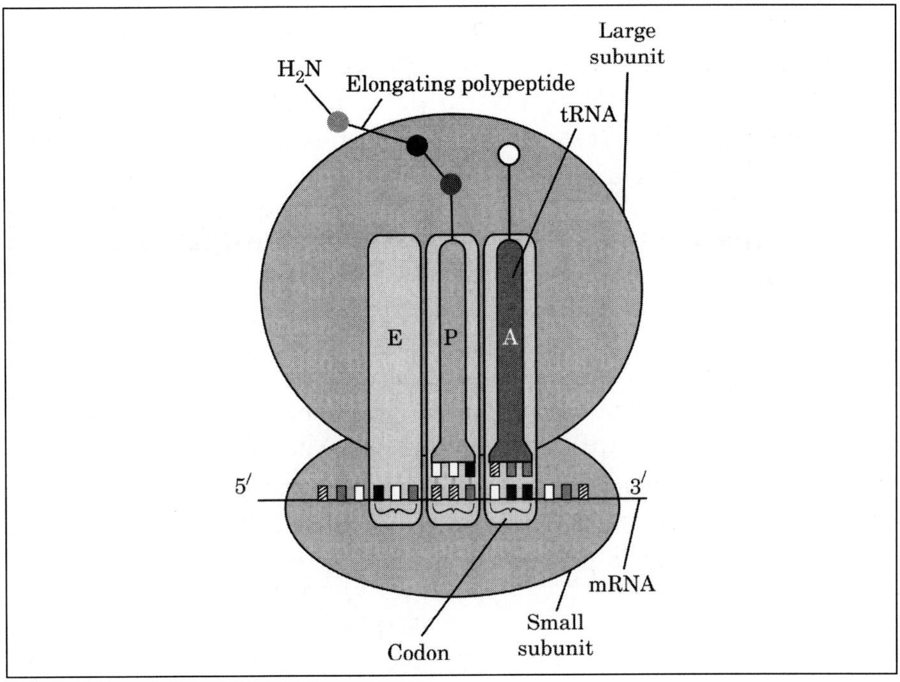

(b) Ribosome unit

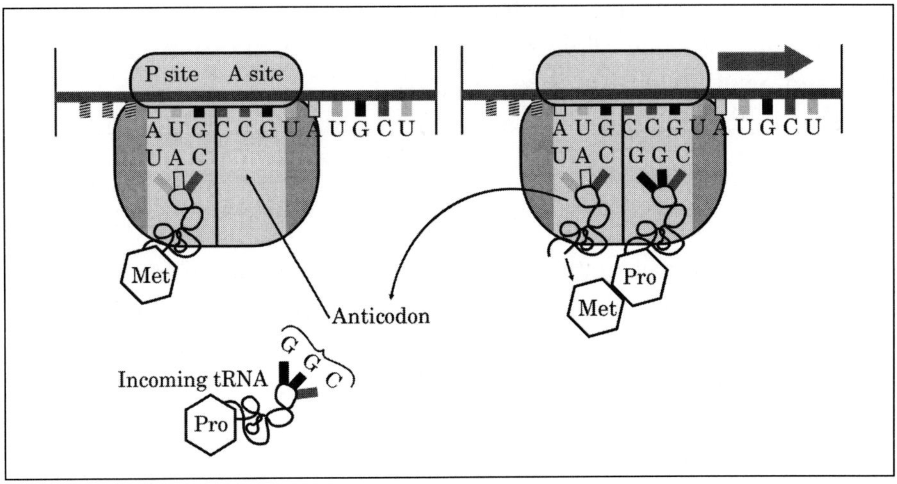

(c) Elongation

Fig. 3.19: The translation process

as the third charged tRNA with amino acid A2 comes to the A site, the tRNA-met-A1 complex moves to site P. Now, A2 is added to the peptide chain, the tRNA shifts to the E site and exits. Repetition of this cycle goes on till the ribosome reaches the termination codon UAA/UAG/ UGA (Fig. 3.19).

With the help of protein release factors the polypeptide synthesized is released in the cell. At a time several ribosomes can attach one after another to the mRNA resulting in simultaneous formation of several polypeptides.

After release, a protein chain undergoes several post translational modifications to be fully functional. Such modifications include removal of the N-terminal met and some more amino acids, addition of chemical groups to the side chains, conjugation with metals and in case of multimeric proteins complexing with other polypeptide chains.

An important point to be remembered is that for the synthesis of a particular protein, translation must start at the correct starting codon. This is because, the sequence of codons is a read-in-frame and a shift in even one nucleotide will lead to a frame shift error in sequence reading. This would lead to the formation of a totally different protein chain as shown below.

...AAC.ACA.CAA.CAA.AAC...	...A.ACA.CAC.AAC.AAA.AC......
asn thr gly gly asn	thr his asn lys

Nature prevents such misreading by ensuring a common starting codon as, for methionine, which can be removed later. Such a fixed starting point ensures that the genetic code is read as a sequence of nonoverlapping triplets.

3.7. REGULATION OF GENE EXPRESSION

The above account describes the processes involved in the flow of information from DNA → RNA → Protein. Major interest lies also in finding how these steps are regulated. There are some genes which are continuously on in all cells and tissues. However, many genes are expressed only in some tissues, at certain times, for sometime and some conditions. Questions as, when, where and how much a gene is expressed, have been active areas of research since specially mid 50s. Much work has been done on prokaryotes and eukaryotic animal systems. In plants, the differential expression of genes of multigene families as those of storage proteins, the photosynthetic enzyme Rubisco, the nitrogen fixing leghaemoglobin protein have been of great interest. There are major differences in the regulation of prokaryotic and eukaryotic genes. In prokaryotes, gene expression is mainly regulated at the transcriptional level, while, in eukaryotes the regulatory controls operate at both transcriptional and post transcriptional levels.

Prokaryotic Gene Expression

In bacteria a protein encoding unit may be polycistronic. Such a unit termed as *operon*, has a row of related structural genes or coding regions or cistrons of a metabolic pathway under the control of a single promoter (Fig. 3.20). An example of transcriptional regulation is the *lac operon* elucidated in 1960s by the pioneering work of Jacob and Monod.

This operon is involved in the transport and breakdown of the milk sugar lactose. It consists of the cistrons *lacZ* for β-galactosidase synthesis, *lacY* encoding permease and *lacA* for acetylase controlled by a single promoter the lac promoter (Fig. 3.20). In this lac promoter, there is an *operator* sequence, to which RNA polymerase can bind and start transcription. However, when lactose is absent, the operon remains inactive as its operator is complexed to a repressor protein encoded elsewhere by a *lacI* gene. In the presence of lactose, the lac operon is activated or induced to metabolize lactose. This happens because lactose itself acts as an inducer by binding to the repressor protein. The lactose-repressor complex detaches from the operator and thus exposes the operator. RNA polymerase can now bind to the operator and initiate transcription of the *lac* genes. This is an example of enzyme induction by the substrate as an inducer.

In another operon the *trp operon*, involved in tryptophan synthesis, tryptophan when in excess, complexes with the repressor and the repressor-trp conjugate blocks trp synthesis by complexing with the operator region of the trp promoter. This is an example of the mechanism of feed back inhibition for the control of intracellular levels of metabolites.

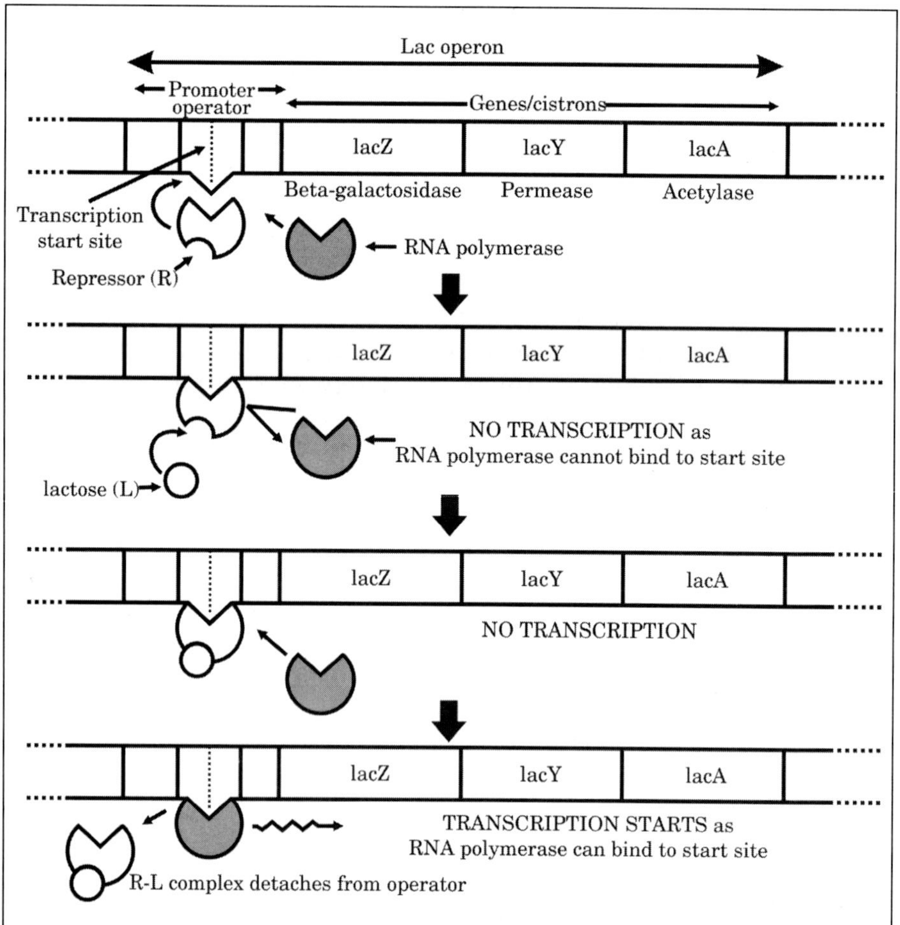

Fig. 3.20: The lac operon

Eukaryotic Gene Expression

Regulation of eukaryotic gene expression is much more complex than in prokaryotes. In prokaryotes the genome is simpler consisting of mainly coding sequences and mostly genes of a metabolic pathway organized together into operons. The genome is also not compartmentalized in a nucleus. In contrast, in eukaryotes the genome is large with genes of even the same metabolism scattered amidst noncoding sequences. Further, although most genes are in the nucleus, the requirements for proteins are in the distant cytoplasm and different organelles. The eukaryotic gene being interrupted, the primary RNA transcript is heterogeneous and needs splicing to form the mRNA. In case of multimeric proteins, the component polypeptides are encoded by different genes and this needs coordinated regulation for synchronous synthesis and assembly. Sometimes in some multimeric proteins as in the case of Rubisco, some genes are in the nucleus and others in the chloroplasts. This entails regulation of transport of the polypeptides formed for subsequent assembly. Therefore, regulation of eukaryotic gene expression occurs at different steps, from the transcription of pre-mRNA to delivery of the final polypeptide product. Thus, the steps under control are, transcription of pre-mRNA, splicing of pre-mRNA to mRNA transcript, modification of mRNA, transport of mRNA from nucleus to cytoplasm and organelles, stability of mRNA, translational control and post translational processing of the polypeptide formed (Fig. 3.21).

Some of the major interventions are the roles of transcription factors, alternate splicing, initiation and repressor factors of translation, transport and post translational processing of the proteins formed.

3.8. CELL CYCLE CHECK POINTS AND CELL DIVISIONS

The chromosome number in all organisms is maintained through generations. This is achieved through checkpoints in cell cycles before and during cell divisions. As is well known, there are two types of cell divisions, mitosis and meiosis. *Mitosis* occurs for proliferation

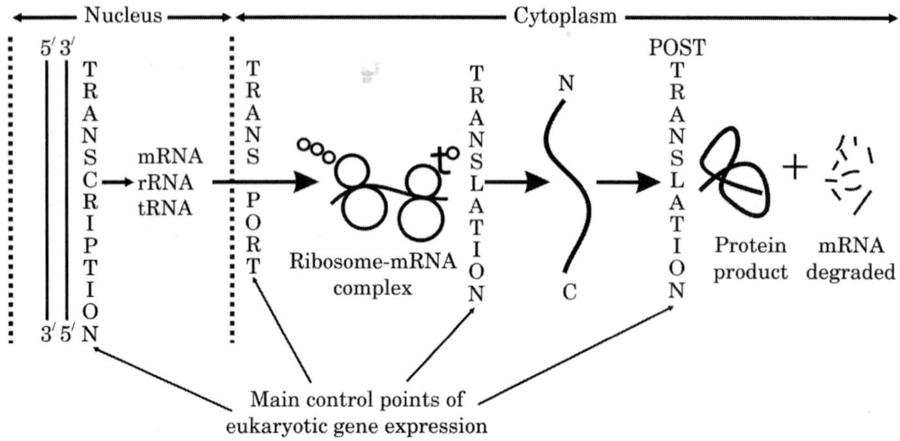

Fig. 3.21: Regulation complexicity of eukaryotic gene expression

of diploid (2n) cells. In this type of division, chromosome number and chromosome characteristics remain same between parent and daughter cells, that is the division is *equational* (Fig. 3.22).

During the production of sex cells as, sperms/pollen and egg cells, a different type of cell division, *meiosis* takes place. During meiosis or *reductional* division, the number of chromosomes is halved in daughter cells. Further, in the first stage of meiosis, homologous chromosomes come close, synapse, exchange parts by crossing over and produce new recombinant daughter chromosomes with variations from parent chromosomes (Fig. 3.23). It may be noted from the Figure that meiosis1 is reductional and meiosis 2 is equational resulting finally in 50:50 ratio of recombinant and parental type chromosomes.

As mentioned above, there are built in check points in the cell cycle for maintaining the integrity of the chromosomes to avoid risky mutations. The main check points are G1/S, G2/M and M. There are many genes involved in the regulation of the cell cycle. These genes produce some *kinases* and *cyclins*. The kinases and cyclins work in combination as, cyclin dependent proteins, the *cdk proteins* for the regulation of the cell cycle. The cdk proteins check, at critical stages or check points of proliferative cells, whether the physiological conditions and quality of DNA are okay or not for the cell to proceed

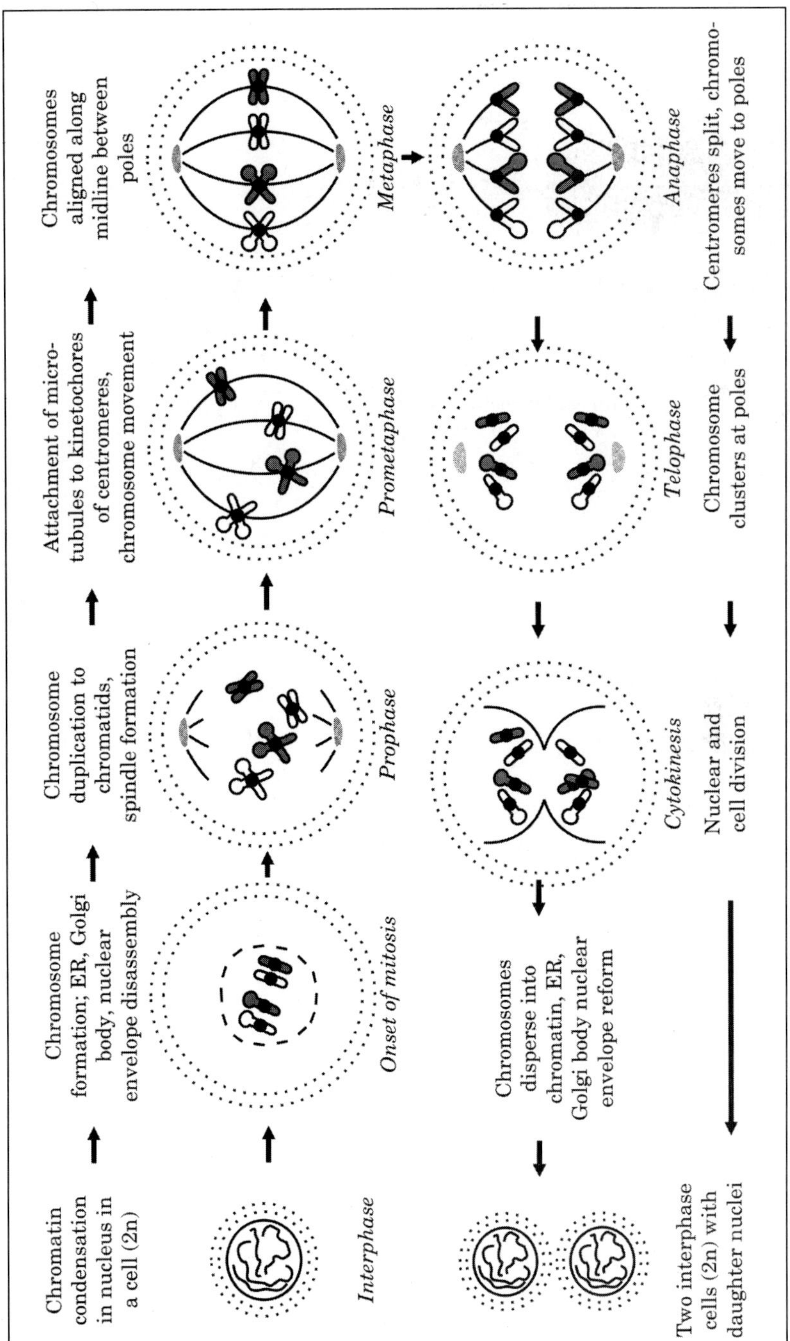

Fig. 3.22: Mitosis

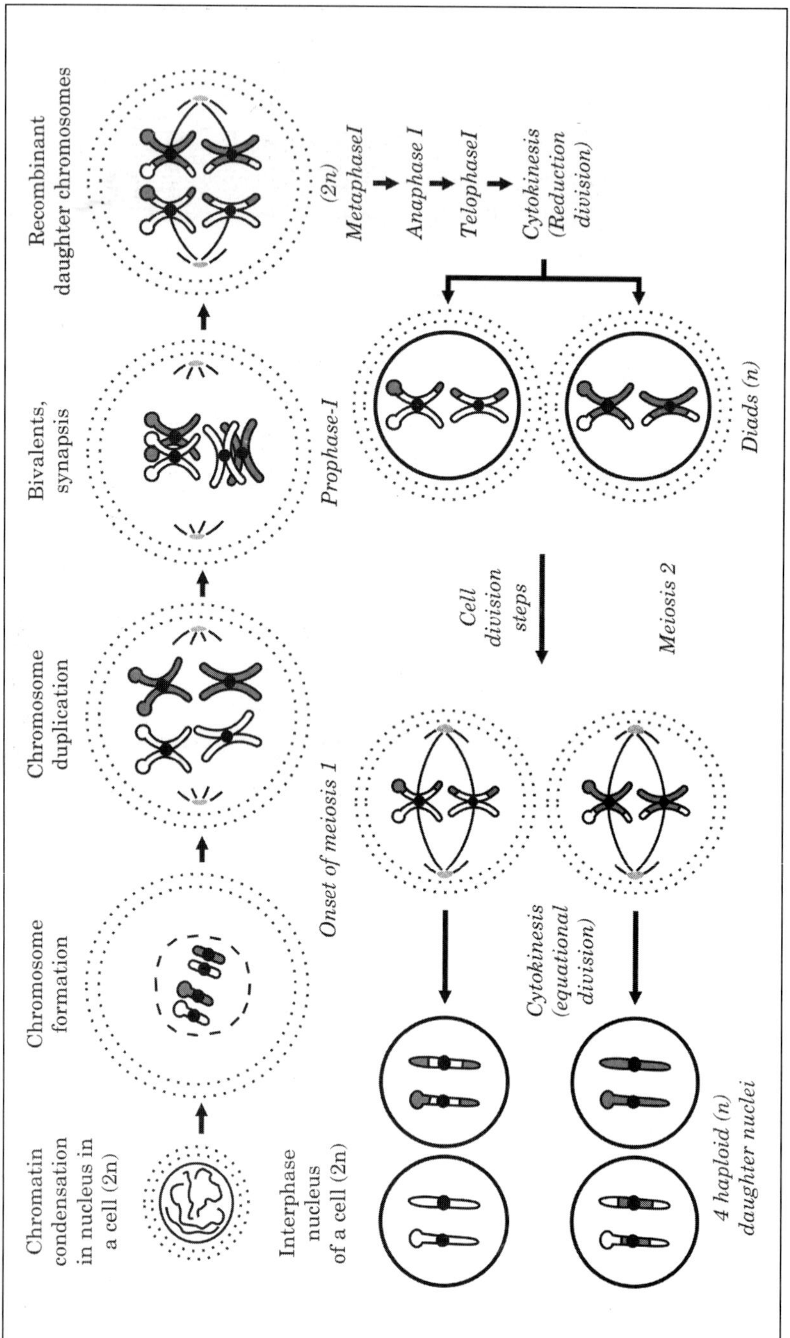

Fig. 3.23: Meiosis

for cytokinesis. Further, there are cellular mechanisms for destroying defective cells also. In humans there is a *p53* gene. Its protein product checks proliferative cells for DNA damage and in case the DNA is damaged, the protein initiates programmed death of the cell or *apoptosis.* In this way, proliferation of a defective cell is prevented. Studies have shown correlations between defective p53 genes and certain forms of cancer. In all organisms, mutations in the genes involved in checks can be deleterious.

3.9. MACROMOLECULAR STRUCTURE AND FUNCTION CORRELATIONSHIPS

In all living organisms, elements have been selected from the mileu of so many elements in the environment to suit certain functional requirements. Similarly, the organic molecules have evolved as building blocks of cellular structures and as functional units to carry out various processes of the living cells. A major role is played by the macromolecules as, nucleic acids, proteins and polysaccharides and high molecular weight lipids. As indicated in the previous sections, nucleic acids and proteins are unique examples of correlations between structure and function. Nucleic acids are mainly for storage and expression of genetic information. However, the transmission of the stored information is enabled by the catalytic activity of so many protein enzymes. A feature to be noted is the underlying principle of economy exhibited during evolution. From mainly four bases, distinct species have evolved with not so distinct differences in genomes. The humans and chimpanzees differ hardly by 2% in DNA base composition (Fig. 3.24 a). The real difference is in the sequence of bases. Similarly, millions of proteins have evolved through differing arrangements of the amino acids. Some proteins serve as structural units and others as enzymes.

Enzymes are organic catalysts and carry out numerous reactions of the cellular metabolism. The compound/s on which an enzyme acts is termed as the substrate and this is either a relatively very small molecule or a small bond. The enzymic reaction proceeds in two steps: (1) formation of an enzyme substrate complex and (2) subsequent formation of the product/s and release of the free enzyme. In all catalysis requiring reactions, the reactants have to be activated or energised before they can react. The energy needed is termed as activation energy. The role of enzyme catalysts is to lower the activation energy, by forming the enzyme substrate complex. The interesting question is, why such a large molecule as a protein, even in the case of a monomeric enzyme, is needed to catalyse reactions of miniscule molecules and bonds? Further, a very small portion of a protein molecule, the active site consisting of few amino acids, is needed for the catalysis. The reason seems to be that, the active site is not a continuous stretch of amino acids, but a cluster of amino acids brought in proximity by folding of the polypeptide. The 3D conformation of the cluster is specific, to match the configuration of the substrate and enable binding of the substrate to the enzyme. This is nature's way of regulating various cellular reactions. When a metabolic step is needed, optimal conditions develop in the cell, as proper pH (negative logarithm of H^+ ion concentration) and temperature. This leads to correct folding of the protein and proper alignment of the amino acids of the active site ensuring formation of a tight enzyme-substrate complex and onset of the reaction. Similarly, when a reaction is not needed, the cellular conditions transform the enzymic protein to an inactive conformation (Fig. 3.24 b). Nature, in such ways, as shifts in the conformational states of enzymes, regulates the cellular pathways of metabolism. The selection of some lipids and proteins as constituents of cell membranes across species, is another marvel of evolution. The membranes basically consist of a phospholipid bilayer with the hydrophilic heads on the aqueous outer and inner sides and the hydrophobic tails in the middle. This foundation layer is interspersed with integral proteins running through the membrane, peripheral proteins and some covalently linked proteins on outer and inner sides (Fig. 3.24 c)). While the middle hydrophobic layer of the tails of phospholipids, prevents en masse entry of water soluble materials, the hydrophilic heads can interact with hydrophilic compounds for various functions. The composition, the types of phospholipids and

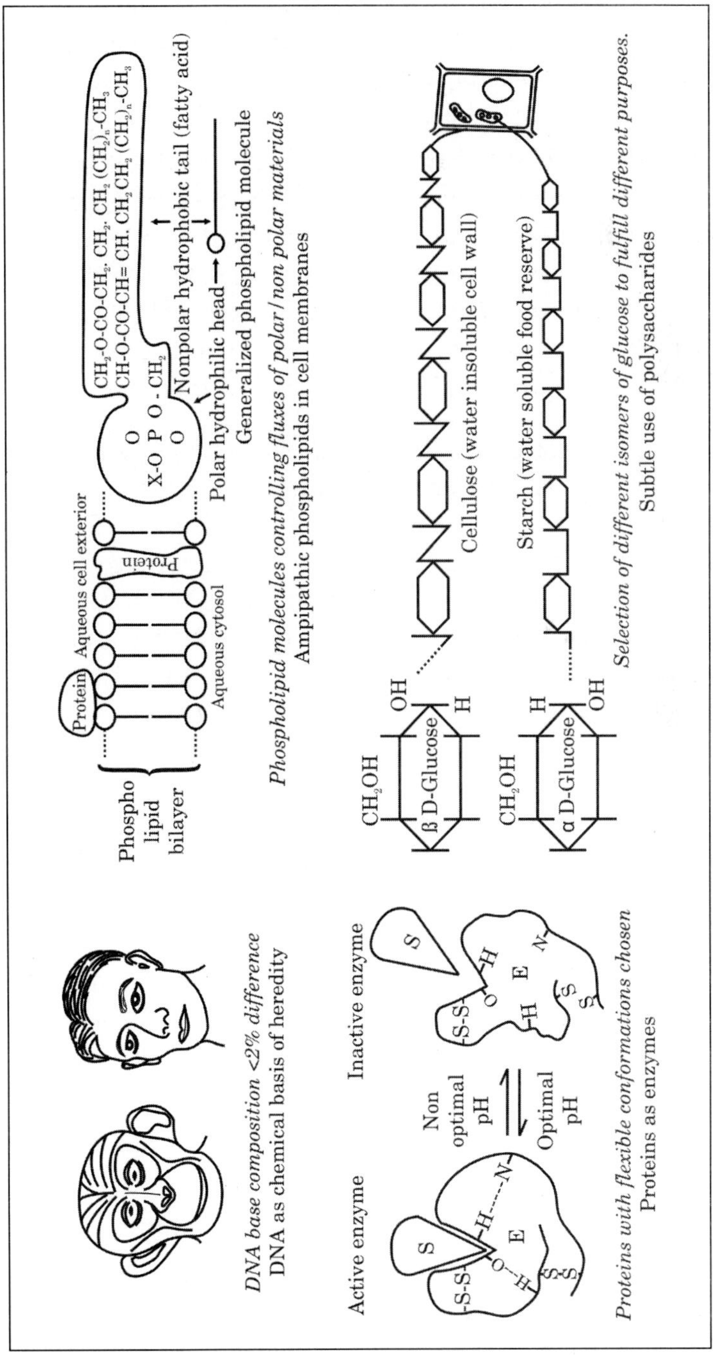

Fig. 3.24: Nature's selection of chemicals for different purposes, small variations big impact

proteins vary with functional requirements of different types of membranes. The simplest and striking example, illustrating the principle of economy in natural evolution, is the use of cellulose as the cell wall in plant cells, for protection of the protoplast and starch as reserve food material. Fibrillar cellulose is insoluble in water and can protect the cell, while, granular starch is the partly soluble easily hydrolyzable food reserve stored in amyloplasts. In both, glucose is the monomer. The only difference is, whereas in starch the glycosidic bond is $\alpha 1 \rightarrow 4$ type, in cellulose it is $\beta 1 \rightarrow 4$ type (Fig. 3.24 d). In the process of molecular evolution, α-D glucose was selected for starch and its isomer, β-D glucose for cellulose, as monomeric units.

4

Basic Techniques of Recombinant DNA Science

SUMMARY OF CONTENTS

Genomic DNA extraction, purification, fractionation; Construction of Vector DNA; DNA of interest recombinant cassette; Amplification of recombinant DNA by cloning; Amplification by Polymerase Chain Reaction; DNA sequencing; DNA microarray technique.

This chapter highlights the principles involved in the protocols for preparing *recombinant DNA*. The task starts with isolation of different types of DNA, specially the source of transgenes or DNAs of interest. The genomic DNA to be studied is broken into fragments mainly by restriction endonuclease enzymes. The fragments obtained are separated by several techniques as electrophoresis to obtain desired short specific DNA segments, such as genes. A target DNA like a gene is ligated or joined to a carrier or vector DNA to make the *recombinant DNA*. To obtain the recombinant DNA in good workable amounts, it is amplified by gene cloning. If the sequence of a gene or DNA fragment is known it can be amplified also by the *polymerase chain reaction*. The first section describes the major steps of recombinant DNA construction and the second is on amplification of the recombinant DNA.

4.1. CONSTRUCTION OF RECOMBINANT DNA

Recombinant DNA is constructed by joining a DNA strand of interest, like a gene, to a vector DNA molecule. Although, many DNA sequences and vectors may be commercially available, it is worthwhile to know the details of the scientific rigour involved in obtaining such molecules. Some of the major steps are described below.

DNA Isolation

For working with DNA, it has to be isolated in pure form first. Sources for total genomic DNA can be eukaryotic tissues, eukaryotic cells and bacterial cells. Further, there is need for isolation of plasmid DNA and bacteriophage DNA. Amongst eukaryotic systems, plant tissues present special challenges because of the presence of interfering chemicals like the phenolics.

Total DNA extraction

The first step is breaking down of cell walls and membranes of eukaryotic or bacterial cell samples. In the case of plants, frozen complex tissue or organ sample is homogenized in a suitable bufferred medium, preferably containing a detergent as sodium dodecyl sulphate (SDS). The homogenate is centrifuged to remove the pellet of cell debris and the supernatant constituting the *cell extract*, is collected. A bacterial cell suspension is centrifuged, the resulting bacterial pellet is collected and treated with a medium usually containing lysozyme and EDTA (ethylene diamine tetra-acetic acid) to breakdown the cell wall. Some SDS is added to disrupt the cell membranes. EDTA also helps in inhibiting DNases, which degrade DNA, by chelating away factors essential for DNase action. After a requisite time the suspension is centrifuged to pellet the cell debris and the supernatant *cell extract* collected. These steps are briefly outlined below:

Eukaryotic tissue + Buffer solution with detergents, DNase inhibitors etc., $\xrightarrow{\textit{Homogenize}}$ Homogenate $\xrightarrow[\textit{collect pellet}]{\textit{Centrifuge and}}$ Cell extract

Prokaryotic microbes	+	Buffer solution as above with cell wall degra-ding enzymes etc.,	*Centrifuge and* ⟶ *collect pellet*	Cell extract

The cell extract, obtained as above, contains a mixture of DNA, RNA, proteins, lipids, carbohydrates and other chemicals. Pure DNA has to be isolated from this mixture mainly by removal of polymeric RNA and proteins. There are several methods of extracting pure DNA from this cell extract. A standard approach is precipitation of proteins, and, enzymic digestion of RNA and residual proteins.

In one technique, the cell extract is treated with phenol or 1:1 mixture of phenol: chloroform which precipitates the proteins. Centrifugation leads to separation of two layers, the aqueous upper and organic lower layer. The precipitated proteins form a layer at the interface of the two layers. The aqueous upper layer with DNA is pipetted out. If the protein content of the cell extract is estimated to be high, it may be useful to treat the extract with some protease enzymes before phenol extraction. After removal of proteins, the extract is treated with ribonuclease enzyme to degrade and remove RNA contaminants. This is followed by another phenol extraction to remove the applied ribonuclease enzyme protein.

DNA is further purified from this protein and RNA free extract, by precipitation with ethanol in the presence of some monovalent cation as Na^+ at a low temperature. The suspension is centrifuged and the DNA precipitate pellet is collected. This pellet is washed with 70% ethanol by resuspending in it, centrifuging and collecting the washed DNA pellet. This purified pellet is redissolved in water to the desired concentration. The main steps are:

Deproteinization and RNA removal:

Cell extract	+	Protein precipitant as phenol	*Centrifuged and upper* ⟶ *aqueous layer collected*	DNA extract containing RNA impurity	*RNA degraded* ⟶ *by RNase and RNase removed with phenol*	Pure DNA extract

Ethanol precipitation:

DNA	+	Ethanol	*Centrifuged and* ⟶ *pellet collected*	DNA precipitate

The amount of DNA extracted is estimated by measuring the absorbance at 260 nm. A_{260} of 1.0 is equivalent to 50 mg of double stranded DNA per ml. Purity of the sample is indicated by the A_{260}/A_{280} ratio, being 1.8 or a little more and not less than 1.8.

In case of plant tissues which contain high amounts of carbohydrates and other compounds as phenolics, a preferred method is the use of a detergent cetyl trimethyl ammonium bromide (CTAB). CTAB forms an insoluble complex with DNA. When this detergent is added to the plant extract, the DNA-CTAB complex is formed as a precipitate and is collected by centrifugation. The collected precipitate is resuspended in 1M NaCl, which dissociates CTAB from DNA. The free pure DNA is recovered by ethanol precipitation as above (Fig. 4.1).

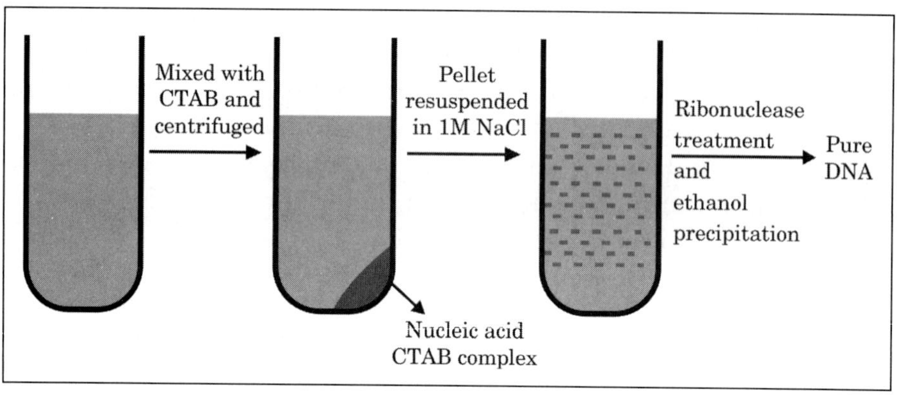

Fig. 4.1: CTAB method of DNA isolation

In another useful protocol, guanidium thiocyanate is used. In a solution of this compound all chemicals except DNA are precipitated or dissolved. Further, DNA binds strongly to silica particles in the presence of guanidium thiocyanate. In practice, guanidium thiocyanate is added to the cell extract and the mixture is poured over a silica column. DNA binds to the silica particles and all other contaminants are washed down by guanidium thiocyanate solution. The retained DNA is eluted out by washing the column with water, which releases the DNA. The DNA obtained is further concentrated and purified by ethanol precipitation.

Plasmid DNA extraction

Plasmid DNA is isolated from cell cultures of plasmid containing bacteria. The main challenge is separation of pure plasmid DNA from the chromosomal DNA. Strategies of separation are based on two differences between the two types of DNA. One approach is based on size difference, the plasmids are much smaller than the chromosomal DNA. The other techniques use difference in conformation, plasmids remain circular or supercoiled while chromosomal DNA forms linear fragments during extraction. The latter methods based on conformation are more reliable.

A cell extract is prepared by first gently degrading bacterial cells with lysozyme and EDTA in the presence of sucrose to yield sphaeroplasts. Lysozyme degrades the cell wall, EDTA prevents activities of specially DNases and sucrose acts as an osmoticum and prevents bursting of the sphaeroplasts, the bacterial cells enveloped only by fragile membranes. Subsequent treatment of the sphaeroplast suspension with some non-ionic detergent as TritonX-100 releases the cellular components. After requisite time the suspension is centrifuged to get the clear relatively plasmid rich lysate. Most of the chromosomal DNA, proteins and RNA sediment in the pellet. The main steps are shown below:

Bacterial suspension +	Buffered medium with isotonic sucrose, DNase inhibitor and Lysozyme	*Incubated* ⟶	Sphaeroplast suspension	*Incubated with* ⟶ *eg., TritonX100, centrifuged and supernatant collected*	Plasmid DNA extract

For obtaining pure plasmid DNA, the lysate can be subjected to alkaline denaturation or preferably ethidium bromide-caesium chloride density gradient centrifugation.

Alkaline denaturation ('miniprep')

The lysate is titrated with sodium hydroxide and SDS to raise the pH to around 12. At this pH the chromosomal DNA strands separate, while the supercoiled plasmid DNA remains intact. Now titrating back to a neutral pH as 7.0 with potassium acetate leads to reassociation of the chromosomal DNA strands into a tangled mass. After centrifugation, the chromosomal DNA mass pellets and the much smaller plasmid DNA remains in the clear supernatant. Plasmid DNA is isolated from this supernatant.

Ethidium bromide density gradient centrifugation

Ethidium bromide intercalates between some base pairs of the normal double helical DNA, leading to partial widening of the interstrand space. This results in lowering of its buoyant density (by about 0.125 g/ml) below the normal of 1.70 g/ml. In contrast, plasmid DNA buoyant density remains about 1.70 g/ml as very few EtBr molecules can intrude into its supercoiled conformation. This phenomenon is utilized in the EtBr-$CsCl_2$ density gradient purification of plasmid DNA. The lysate is layered over EtBr containing caesium chloride solution in a centrifuge tube and ultracentrifuged. After due period, different macromolecules equilibriate, densitywise, with different regions of the $CsCl_2$ density gradient, proteins being of lowest density at the top, followed bottomwards by chromosomal DNA above 1.70 g/ml region, plasmid DNA at 1.70 g/ml region and the RNA pellet at the bottom. The DNA bands are visualized by UV exposure and the plasmid DNA band extracted. Pure plasmid DNA is obtained by removing EtBr with n-butanol extraction and $CsCl_2$ by dialysis (Fig. 4.2).

Bacteriophage DNA extraction

Bacteriophage viral DNA exist as free particles in the extracellular medium of infected bacteria. Therefore, the phage particles are pelleted by ultracentrifugation and the phage DNA extracted after deproteinization. However, the protocols for obtaining sufficient number of phage particles in the extracellular medium varies with lysogenic *e.g.*, lambda (λ) and lytic *e.g.*, M13 types of phages.

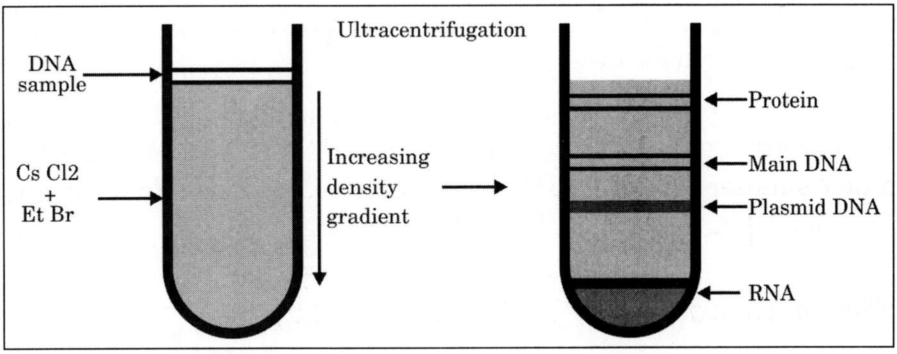

Fig. 4.2: Ethidium bromide separation of plasmid DNA

Lysogenic phages (e.g., λ phage)

Lysogenic phage DNA, on entering the bacterial cell integrates into the bacterial DNA and remains as a quiescent phage. Only due to some stimulus, the phage DNA gets released into the cell, forms phage particles, causes lysis of the cell and discharges into the medium. The trick is providing the stimulus to shift the phage to the lytic cycle. Usually a strain having a mutation in a gene is used. A popular strain has the temperature sensitive *ts* mutation in one of the integration regulating genes, the *cI* gene. The *cIts* gene keeps phage DNA integrated in the chromosomal DNA at 30°C, but, at 42°C becomes dysfunctional resulting in the onset of the lytic phase with release of the phage DNA, formation of λ phage particles and their release into the culture medium. Care is taken to obtain high yields, higher than 10^{10} / ml, for getting workable amounts of DNA. After sufficient particles form, the medium is ultracentrifuged to remove the cell debris as pellet and collection of the phage containing supernatant. The phages are precipitated with polyethylene glycol (PEG), the PEG phage precipitate collected by centrifugation and resuspended in a small volume of a suitable buffer.

This preparation is usually subjected to $CsCl_2$ density gradient centrifugation and the lambda particle layer in the approx. 1.50 buoyant density zone pipetted out. The phage DNA is collected by phenol deproteinization, ethanol precipitation and washing. The basic steps are shown below.

Lysogenic *Cl* λ phage integrated with bacterial DNA at 30°C —(*heated to* *42°C*)→ Onset of lysis and phage DNA released in medium —(*ultracentrifuged* *and supernatant collected*)→ Phage particle suspension

Phage particle suspension + PEG —(*Centrifuged* *and precipitate collected*)→ Phage-PEG in suitable buffer —(*Gradient ultracentrifugation, deproteinization, ethanol precipitation and washing*)→ Phage DNA

Lytic phage (e.g., M13)

Extraction of such phage DNA is simpler. An infected bacterial culture is grown. After requisite growth, the bacteria and debris are removed by centrifugation, the phage particles in the supernatant precipitated by PEG. The PEG phage precipitate is collected by centrifugation, deproteinized with phenol, the phage DNA is precipitated and washed with ethanol.

4.2. DNA FRACTIONATION BY GEL ELECTROPHORESIS

Each of the DNA types obtained by the methods mentioned above are mixtures of different DNAs. For obtaining specific pure DNA, the total DNA sample has to be subjected to fractionation procedures. DNA at neutral pH is negatively charged and when placed between electrodes moves to the positive electrode or anode. Agarose or polyacrylamide gel electrophoresis is commonly used for separations of the fractions. Separations depend on the type of gel, gel concentration and thickness of the slab. Thus, 1% agarose separates DNA fragments from 200 bp to 2 kb, 0.5% agarose gel of 0.5 cm slab thickness, separates from 1 to 30 kb while a 0.3 mm, 40% polyacrylamide gel separates DNA strands ranging from 1 to 300 bp.

An electrophoresis apparatus consists of a gel layer on a glass or plastic sheet, the two edges of which rest on tanks each containing a buffer. The gel is layered on the sheet and allowed to set. Each tank has an electrode connected to the gel as also to a DC power supply. The total DNA sample is dissolved in a neutral buffer, containing a dense liquid as glycerol and a dye and loaded in a well near the negative electrode. The dye indicates progress of electrophoresis and glycerol ensures settling of DNA in the bottom of the well. In an adjoining well, a mixture of DNA in a similar medium of known sizes, the standard markers, is also pipetted. The current is put on for the electrophoretic run. DNA fractions being negatively charged move towards the positive electrode, the rate of movement depending on the size. Smaller molecules move faster than the larger molecules. After the run the bands are visualized with the help of a stain as ethidium bromide which fluoresces under UV irradiation. In the case of radio isotope labeled starting sample, visualization can be done by autoradiography also. The sizes are estimated by comparing with the standard bands running parallel to the sample line and calculated from the relationship d $\propto$ $\log_{10}$ of the molecular weight of the fragment. The steps of the protocol are shown in Fig. 4.3. A more detailed account of some electrophoretic protocols and detection of the separated bands by various methods are given in a later section on 'fractionation and detection of macromolecules'.

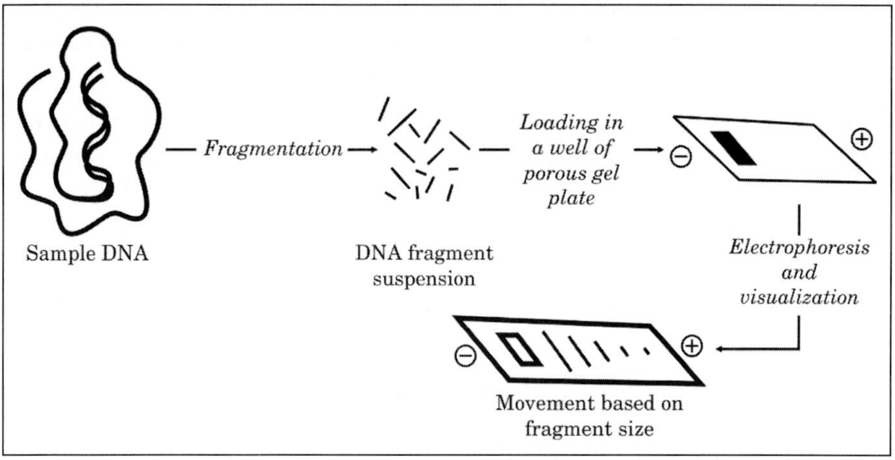

Fig. 4.3: Gel electrophoresis

4.3. ISOLATION OF SPECIFIC DNA MOLECULES (*E.G.,* A GENE)

The DNA molecules isolated by above steps are very large, consisting of many genes or coding sequences and noncoding sequences. For studying individual genes, it is necessary to break such large DNA molecules, for instance a chromosomal DNA, into smaller units *e.g.*, into fragments, with some strands representing single genes. This is done with the help of *restriction endonuclease enzymes,* described later. The sample of a large DNA molecule is broken with *restriction enzymes*, the products separated by appropriate *gel electrophoresis* and specific molecules identified by *hybridization with labeled complementary probes*. Following this, the pure specific DNA is isolated. However, the amount is very small. Large workable amounts of this DNA is obtained by amplification through the processes of *gene cloning* and/or PCR *(polymerase chain reaction).*

Cutting with Restriction Enzymes to Obtain DNA Fragments

Certain strains of bacteria are immune to infection by bacteriophages. This is because these strains have some endonucleases which *restrict* the invading viruses by degrading their DNA. Such enzymes which break down the infecting viral DNA are called *restriction enzymes*. The interesting feature is that while these enzymes can attack and degrade the alien DNA, the chromosomal and extrachromosomal DNAs of the bacterium remain unaffected. This is because, the bacteria have unique restriction-modification systems. Each such system consists of a *restriction* enzyme which recognizes a specific base sequence in the invading DNA *e.g.,* phage DNA and a *methylase*, which protects the host bacterial DNA by methylating same sequences present in it. As restriction enzymes cut DNA at specific sites they are used extensively for obtaining specific DNA fragments. The name of each restriction enzyme is derived from the name of its source bacterium and specificities. The binomial nomenclature system has been adapted. The *first letter* in uppercase is for genus, following *two letters* in lower case for species and the order of characterization in *roman letters*. Thus, three restriction endonucleases of different specificities, isolated from the bacterium *Haemophilus aegypticus* are named, *HaeI, HaeII and HaeIII.* There are many restriction enzymes and they have been classified into three groups, Type I, Type II and Type III. For precise cleaving of DNA, the Type II restriction endonucleases are used. For cutting the DNA, the enzyme attaches to a specific site, a sequence of nucleotides, which it recognizes, the *recognition site*. The length, composition and the exact site of excision of the recognition sequences of different restriction enzymes vary as shown in the following Table 4.1.

Table 4.1: Examples of some restriction enzymes

Enzyme	***Sequence***	***Base number***	***Cut end***	***Cut frequency (1 in 4*)***
Sau3AI	5′-/GATC-3′ 3′-CTAG/-5′	4	5′Sticky	0.25 kb
AluI	5′-AG/CT-3′ 3′-TC/GA-5′	4	Blunt	0.25 kb
EcoRI	5′-G/AATTC-3′ 3′-CTTAA/G-5′	6	5′Sticky	4.0 kb
XmaI	5′-C/CCGGG-3′ 3′-GGGCC/C-5′	6	5′Sticky	4.0 kb
SmaI	5′-CCC/GGG-3′ 3′-GGG/CCC-5′	6	Blunt	4.0 kb
NotI	5′-GC/GCGCGC-3′ 3′-CGCGCG/CG-5′	8	5′Sticky	65.0 kb

Several interesting features of the restriction enzymes emerge by a perusal of the above table. The recognition sites are palindromic, the nucleotide sequence being same in both strands in the same $5' \rightarrow 3'$ direction. Further, the length of the recognition site determines the possible frequency of occurrence of the attachment sites in the substrate template DNA. For instance, in case of a quadramer site (4N) the site will occur approximately 1 in 4^4 bases or every about 250 bp and in case of a hexamer, 1 in 4^6 bases or every approximately 4000 bases. This helps in deciding which RE to use for a DNA of known length. However, this calculation is based on the presumption that all bases occur in equal proportion (G+C = A+T) and their distribution is random, which rarely happens. Therefore, the sites for restriction enzymes have to be found experimentally by *restriction mapping.*

Cutting by some enzymes yields products with sticky cohesive ends and others release products with blunt ends. Products with complementary sticky ends anneal easily when in close proximity. This feature is utilized extensively for joining different DNA molecules cut by same or similar restriction enzymes. Products with sticky and blunt ends are shown below.

		Sticky ends		
	EcoRI			
5'..ATTGCCATT***G/AATTC***..TTGCAA....3'	⟶	5'..ATTGCCATT***G***		***AATTC***..TTGCAA....3'
3..TAACGGTAA***CTTAA/G***..AACGTT....5'		3..TAACGGTAA***CTTAA***	+	***G***..AACGTT...5'

		Blunt ends		
	SmaI			
5'..ATTGC***CCC/GGG***AAGCGC..3'	⟶	5/..ATTGC***CCC***	+	***GGG***AAGCGC..3'
3'..TAACG***GGG/CCC***TTCGCG...5'		3'..TAACG***GGG***		***CCC***TTCGCG...5'

The table further shows that *XmaI* and *SmaI* have similar recognition sequences and are therefore, *isoschizomers.* These enzymes however differ in their sites of action, *XmaI* yielding sticky end products and *SmaI* releasing blunt end products.

Restriction enzymes as *NotI* which recognize long restriction sites are mainly used in manipulations as genome sequencing.

Restriction Mapping

The DNA molecule of interest is digested with several restriction enzymes, with one enzyme at a time as also with 2 or more enzymes at a time. In each case, the fragments formed are separated by gel electrophoresis, and their sizes determined. By comparison of the fragment sizes in single and mixed digestions, the positions of the sites of different restriction enzymes are deduced and the restriction map constructed. This is illustrated through a simple example given below:

Lambda DNA, of 48.5 kb length, is digested with the enzymes X*ba*I and X*ho*I individually and in combination. The products are

Enzymes	*Fragment number*	*Fragment sizes, kb*	*Total, kb*
*Xba*I	2	24.0, 24.5	48.5
*Xho*I	2	15.0, 33.5	48.5
*Xba*I + *Xho*I	3	09.0, 15.0, 24.5	48.5

The above table shows that *Xba*I produces two fragments of nearly equal sizes 24.0 and 24.5 kb. *Xho*I produces one small 15.0 kb fragment and one larger 33.5 kb fragment. The question is, to which of the two *Xba*I fragments the smaller 15 kb *Xho*I oligonucleotide belongs to. The mixed digest products provide the answer. In the mixed digest, the 24.5 *Xba*I fragment is intact and therefore, the *Xho*I 15 kb fragment is a part of the 24.0 kb *Xba*I product. Therefore, the restriction map for *Xba*I and *Xho*I restriction enzymes would be:

Lambda DNA 5/.. 3/48.5 kb
48.5 kb

*Xba*I .... 48.5 kb
24 kb 24.5 kb

*Xho*I .... 48.5 kb
15 kb ?? 33.5 kb

*Xba*I + *Xho*I .... 48.5 kb
9 kb 15 kb 24.5 kb

Lambda DNA Xbal Xhol
Restriction Map 5′ ...............................I.....................I..3′
48.5 kb

Restriction maps like the one shown above, are necessary for isolating a sequence of interest, like a gene, from a sample DNA.

4.4. RECOMBINANT DNA CONSTRUCTION BY DNA-INSERT: VECTOR-DNA LIGATION

After isolation of a gene or a DNA sequence of interest, it has to be amplified to workable amounts. One of the methods is gene cloning. For the important tool of gene cloning, it is necessary to join the DNA to be cloned to a vector DNA molecule. *The complex of ligated gene or a DNA fragment and vector DNA is the Recombinant DNA.* The vector molecule is a carrier molecule that carries the DNA insert attached to it into a bacterial cell or even to cells across species barriers, for instance, a bacterial DNA into plant DNA.

Joining of two different DNA molecules is done with the help of DNA ligases (Fig. 4.4). The most commonly used ligase is the T_4 DNA ligase, isolated from T_4 bacteriophage infected *E. coli* bacteria. DNA ligase repairs a discontinuity, that is absence of a phosphodiester bond between two nucleotides, in a DNA strand. The enzyme reacts with ATP to form an enzyme-AMP covalent complex, which binds to the discontinuity site and restores the phosphodiester bond.

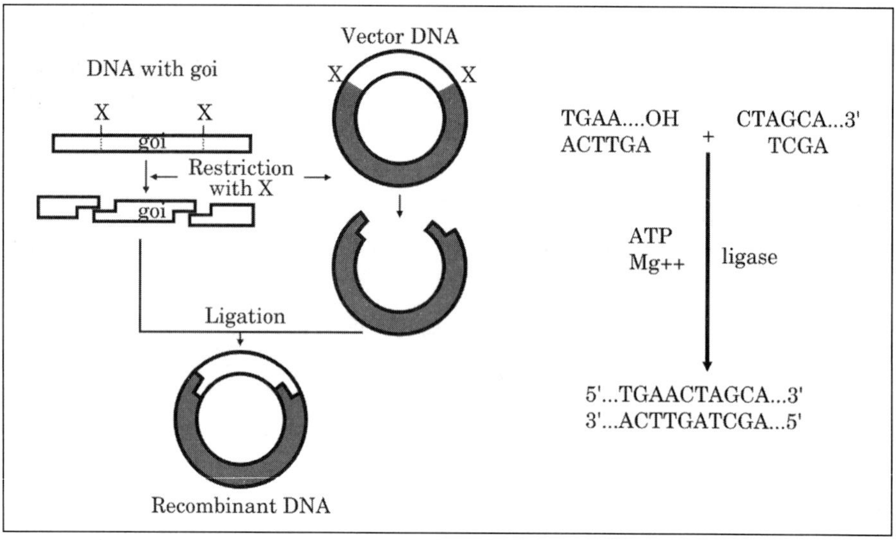

Fig. 4.4: Ligation, (a) Recombinant DNA, (b) Action of T4 DNA ligase (X = a restriction enzyme)

Sticky End Ligation

Two DNA molecules, say an insert gene DNA and a vector DNA, each with sticky ends are easiest to ligate and form a recombinant DNA molecule. Complementary sticky ends of the insert and the vector DNA molecules anneal through base pairing, thus providing a platform for the ligase to attach and conduct ligation.

There are however, complications like, intramolecular ligation such as more of vector-vector or gene-gene ligation, rather than gene-vector intermolecular ligation. One major condition, to ensure maximum intermolecular ligation is to have an optimum insert:vector concentration or molar ratio. The normal molar ratio of insert:vector = 3:1 ratio or a size based ratio of W_v /S_v : Wi/Si = 1:1, where W_v, S_v are the weight and size of the vector and W_i, S_i are weight and size of the insert. Such a ratio ensures excess of insert DNA relative to vector DNA.

Blunt End Ligation

Recombinant DNA formation with blunt-ended DNA molecules is difficult as ligation proceeds at a much slower rate. This is because the blunt ends of the insert and the vector DNA come in proximity through chance associations and not complementarity. Therefore, very high concentrations of both types of molecules are needed to enhance the probability of chance associations. Yet the ligation will not be efficient as, with relatively large insert and vector DNA molecules it becomes difficult to achieve the optimal molar ratio. Efficient combination of blunt-ended DNA can however be achieved by the use of *linkers, adaptors or homopolymers*. These are synthetic oligonucleotides for attaching to the ends of the DNAs to be joined.

Linkers

These are short double stranded blunt-ended oligonucleotides designed to contain a site for a restriction enzyme (RE). The DNA to be cloned is mixed with a high concentration of the selected linker molecules and a ligase. Despite being blunt-ended, the high concentration of the low molecular weight linker, allows for linker-sample DNA joining. Treatment with a restriction enzyme, the recogniton site for which exists in the linker, results in the production of sample DNA-linker complexes with sticky ends (shown below). Care is taken, not to include any restriction site in the linker that occurs in the sample DNA.

```
Sample DNAs                             RE              Blunt ended products
5'--*************I************3' ──────────────▶ 5'*************--      --************3'
3'--*************I************5'                 3'*************--      --************5'

                                   Linker with
Blunt end products                 EcoRI site      Ligation  Sample DNA with linker
5'**********-    -*********3'  +  5'----G/AATTC----3' ──▶ 5'*********---G/AATTC---*******3'
5'**********-    -*********3'     3'----CTTAA/G----5'     3'*********---CTTAA/G---*******5'

Sample DNA with linker              EcoRI        Blunt end products with sticky ends
5'*******---G/AATTC---******3' ──────────▶ 5' *******---G     +    AATTC---******3'
3'*******----CTTAA/G---******5'            3' *******----CTTAA          G---******5'
```

Adaptors

Adaptors are short synthetic double stranded oligonucleotides with one blunt end and the other sticky end. The blunt end of the adaptor is ligated to the DNA to be cloned to produce a new complex with a sticky end. To prevent adaptor-adaptor ligation after complementary annealing, the 5′ terminus of the sticky end of the adaptor is dephosphorylated, while the blunt end is retained in the natural form with 5′P and 3′OH termini. After attachment of the adaptor, the DNA fragment is treated with polynucleotide kinase to phosphorylate the 5′ terminus of the sticky end of the attached adaptor (shown below). Now this adaptor-DNA complex can be inserted into a vector.

```
        Sample DNAs                    RE              Blunt ended products
5'--*************I************3' ──────────────▶ 5'*************--      --************3'
3'--*************I************5'                 3'*************--      --************5'

Synthetic oligonucleotides       EcoRI          5' phosphorylated adaptors (in bold)
5'*******G/AATTC****3' ─────────────────────▶ 5'******G              5 P-AATTC****3'
3'*******CTTAA/G****5'                         3'******CTTAA-P5'              G****5'

Dephosphorylated adaptors
5'OH-AATTC****3'   +   3'******CTTAA-OH5'

Sample DNA with rephosphorylated adaptors as sticky ends
5'*************--*******G                      5'P-AATTC--**********3'
3'*************--*******CTTAA-P5'                      G--**********5'
```

Homopolymer tailing

In this method, sticky ends are produced by adding homopolymers to the 3′-OH termini of a double-stranded DNA molecule. A homopolymer is an oligonucleotide containing the same deoxynucleotide, poly N, for instance poly G or poly A. The method is to treat the sample DNA with the enzyme terminal deoxynucleotide transferase, in the presence of one deoxynucleotide, like G or C or A or T.

Usually, the two DNA molecules to be combined are tailed with complementary bases. For instance, to an insert DNA one may add a poly T tail and to the vector a poly A tail. The length is of about 20 bases. The insert and the vector having complementary tails anneal and can now be ligated with DNA ligase as shown below.

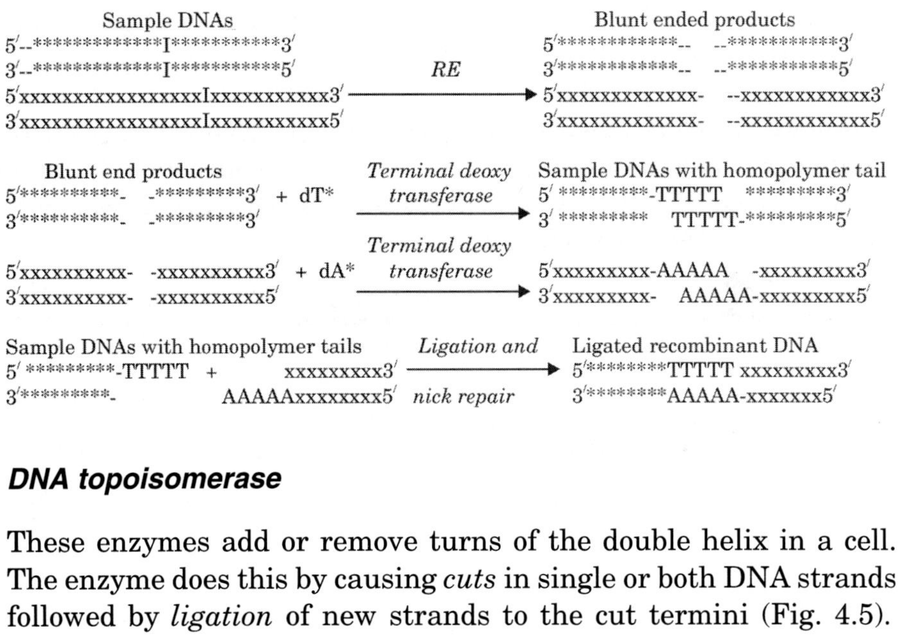

DNA topoisomerase

These enzymes add or remove turns of the double helix in a cell. The enzyme does this by causing *cuts* in single or both DNA strands followed by *ligation* of new strands to the cut termini (Fig. 4.5).

For making plasmid-insert recombinant DNA with this enzyme, a plasmid with a recognition sequence is treated with a specific topoisomerase. The enzyme attaches to the recognition site cuts and linearizes the plasmid molecule. The product with the enzyme bound to it has $3'$-P and $5'$-OH termini at the blunt ends. Topoisomerase can only join DNAs with $3'$-P and $5'$-OH containing blunt ends. The DNA to be cloned has blunt ends with normal $5'$-P and $3'$-OH termini.

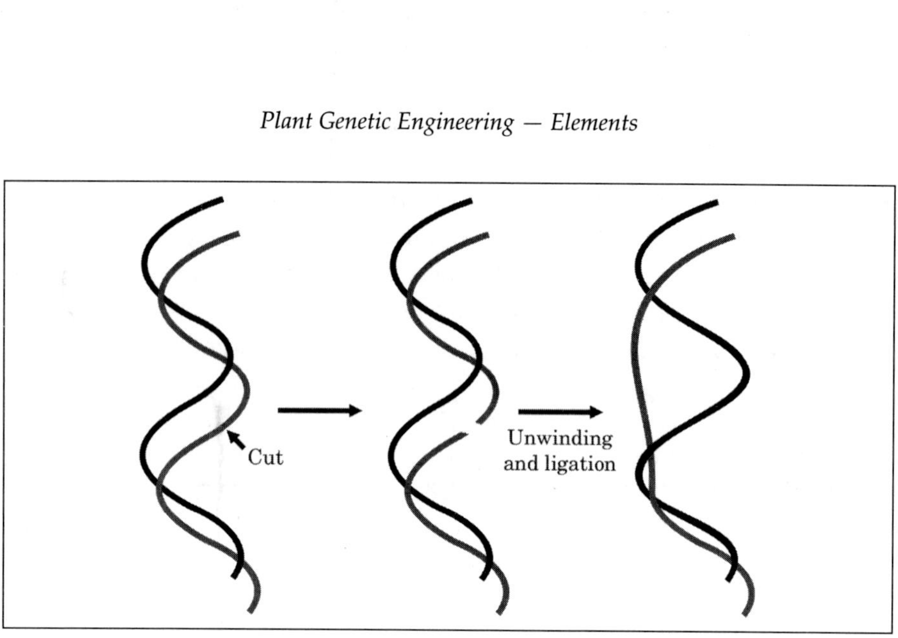

Fig. 4.5: DNA topoisomerase

Therefore, for ligation with topoisomerase, this DNA is treated with alkaline phosphatase to remove the 5′ terminus P. Such phosphatased DNA is mixed with the vaccinia plasmid-topoisomerase preparation for ligation to proceed (shown below).

Plasmid vector DNA	Topoisomerase	Plasmid vector DNA Blunt ends	
5′**********************3	⟶	5′************3′P	OH5′**********3′
3′**********************5′		3′************5′OH	P3′**********5′

Insert DNA blunt end products		Alkaline	Dephosphorylated insert blunt end products	
P-5′xxxxxxxx-3′OH	P-5′xxxxxxxx3′-OH	⟶	OH-5′xxxxxxxxx3′-OH	OH-5′xxxxxxxxxx3′-OH
OH-3′xxxxxxxx-5′-P	OH-3′xxxxxxxx5′-P	phosphatase	OH-3′xxxxxxxxx5′-OH	OH-3′xxxxxxxxxx3′-OH

Vector DNA		Insert DNA	Topoisomerase	Vector-Insert Recombinant DNA
5′***********3′P	+	OH-5′xxxxxxxxxxx3′-OH	⟶	5′***********---------xxxxxxxxxx3′-OH
3′***********5′OH		OH-3′xxxxxxxxxxx5′-OH		3′***********OH OH-xxxxxxxxxx5′-OH

4.5. AMPLIFICATION OF RECOMBINANT DNA BY GENE CLONING

Gene cloning helps in increasing the amount of recombinant DNA constructed, from nanograms to say milligrams. Further, it helps in getting a pure sample of the desired recombinant DNA free from contaminants as vecto-vector, unligated vector or wrong insert-vector DNAs. The vector or carrier molecule to which the gene DNA is ligated, is usually a derivative of a suitable bacterial plasmid DNA or a viral bacteriophage DNA. Both plasmid DNA and phage DNA selected as vectors, are capable of multiplying in the bacterial cell independent of the chromosomal DNA. Therefore, the gene-vector recombinant DNA when introduced in bacteria also replicates and multiplies, leading to massive amplification of the gene DNA. Successful introduction of plasmid DNA in bacteria is termed as *transformation* and the products as *transformed* cells. Phage DNA insertion is termed as *transfection*. It may be noted that transformation requires uptake of DNA molecules by bacteria. Some bacteria have natural ability of taking up DNA and are termed as *competent*. Strains of *Pneumococcus* are naturally *competent*. These strains of *Pneumococcus* were used by Griffith in 1920s and later, Avery and McLeod for transformation of an avirulent strain to a virulent strain. The transformation was because of natural uptake of DNA from virulent strain into avirulent cells. However, all bacteria, specially *E. coli* strains, are not naturally competent and need an ice cold salt wash and subsequent 'heat' shock to become *competent*.

There are some important requirements to be fulfilled in vector designing (Fig. 4.6). One is inclusion of a sequence for independent replication of the plasmid, independent of chromosomal DNA control. The ori sequence (origin of replication) ensures that. Further, *marker* sequences for selection of transformed cells are necessary as also

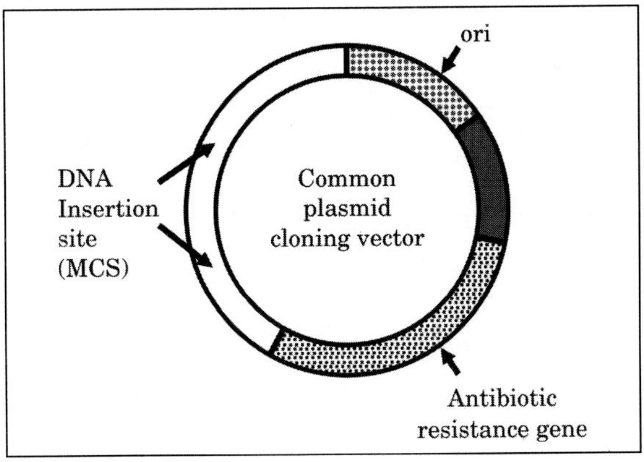

Fig. 4.6: Essential components of a vector DNA molecule

several *restriction sites* which help in insertion of the DNA to be cloned. Considerations of number of copies formed and host range are also important. Thus, the ColE1 derived ori sequence in a plasmid pBR322 allows for about 20 copies to be formed while its slightly modified form in pUC18 induces around 100 replications. This ori has, however, a narrow host range being able to function in *E. coli* and some related species only. Broad range vectors are derived from R(resistance conferring) or F(fertility conferring) plasmids. Vectors have been described in detail in Chapter 5.

Gene Cloning by Bacterial Transformation with Recombinant Plasmid DNA

The protocols involve the transformation process followed by selection of the transformed bacteria.

Transformation

Transformation involves introduction of a recombinant plasmid vector into a bacterial cell (Fig. 4.7).

The most commonly used bacterium is *Escherichia coli*. Transformation is not a very efficient process as *E. coli* cells are

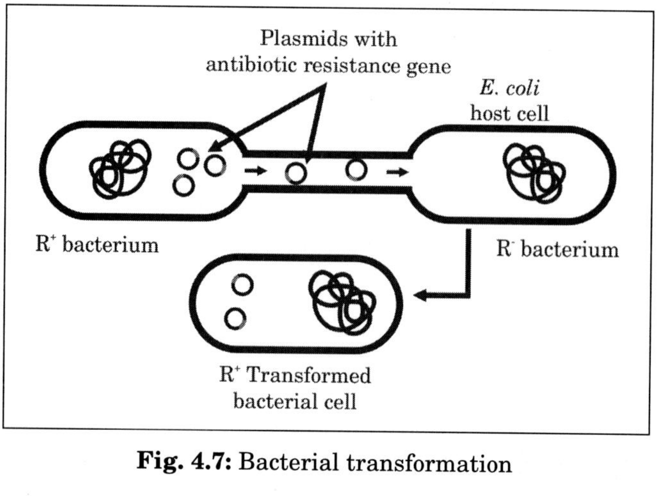

Fig. 4.7: Bacterial transformation

normally not *competent* to take up of DNA molecules from the ambient medium. To confer *competence*, the host bacterial cells are first treated with ice cold $CaCl_2$, recombinant DNA is added and the mixture is given a brief heat shock by raising the temperature to 42°C. $CaCl_2$ treatment ensures binding of DNA to the bacterial cell walls and the heat shock, penetration of the DNA into the cells. The mixture is incubated for sometime before plating on solid agar medium, laid in petriplates. After about 24h, colonies of bacteria appear on the agar plates. Since each colony develops from a single bacterium and has the same genetic makeup, it is called a *clone*. As mentioned earlier, the transformation efficiency is quite low (10^9 transformants/microgram plasmid DNA) and the colonies of transformed bacteria containing the gene-vector recombinant DNA have to be identified and selected. Advantage is taken of a natural property of the plasmids chosen for gene cloning. Most plasmids carry some genes for antibiotic resistance and confer this property to the transformed bacterial cells. Thus, a common plasmid pBR322 has genes for both ampicillin and tetracycline resistance. When this plasmid enters *E. coli* cells, otherwise susceptible to these antibiotics ($amp^S tet^S$), the daughter cells get transformed and become ampicillin and tetracycline resistant ($amp^R tet^R$). When susceptible cells ($amp^S tet^S$) are plated on antibiotic containing agar, they get killed and colonies are not formed. In contrast, the resistant bacteria ($amp^R tet^R$) survive and form colonies (Fig. 4.8). This ability to distinguish transformed and nontransformed cells forms the basis of selection of the transformed cells.

Selection of transformed cells (recombinant selection)

There are two main methods, selection by insertional inactivation of antibiotic resistance gene and insertional inactivation of *lacZ* gene.

Recombinant selection by insertional inactivation of antibiotic resistance gene

In practice, $amp^S tet^S$ *E. coli* cells are incubated with the recombinant DNA construct solution. The uniqueness of the technique is that the recombinant DNA is made by inserting the gene of interest (goi) in the tetracycline gene segment, thereby inactivating it. It must be remembered that the recombinant DNA solution contains, besides the insert-vector DNA recombinant DNA, unligated reformed vector and vector-vector DNA molecules also. In the reformed vector the genes for both ampicillin and tetracycline resistance are present. So transformation by these DNAs confer resistance to both antibiotics, the transformed cells becoming $amp^{R}tet^{R}$. On the other hand, bacteria with gene-vector recombinant DNA are resistant only to ampicillin and sensitive to tetracycline. This is because, as mentioned above, the insertion of the gene DNA into the vector molecule is designed to occur in a gap created in the tetracycline resistance gene segment. Such insertion disrupts the tetracycline resistance gene cluster leading to its inactivation and the gene-vector recombinant having the $amp^{R}tet^{S}$ configuration. This technique is termed as *insertional inactivation* of a gene (Fig. 4.8).

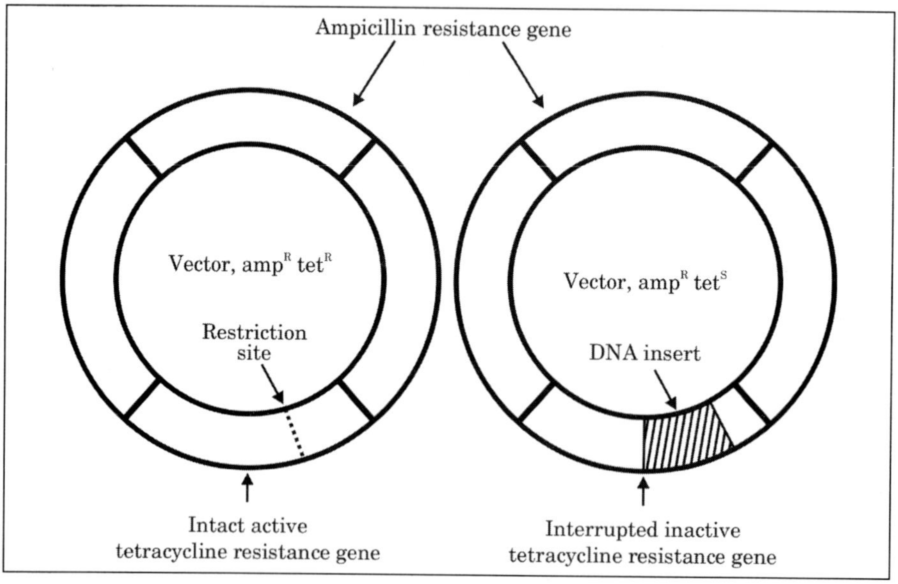

Fig. 4.8: Insertional inactivation of an antibiotic resistance gene for selection of transformed recombinant *E. coli* (R = resistant, S = susceptible)

The cells transformed by the recombinant DNA with insertional inactivation of the tetracycline resistance gene, will be $amp^R tet^S$, that is resistant to ampicillin and susceptible to tetracycline. For selection of cells with gene-vector DNA, the recombinant DNA-bacterial cell incubation mixture is first plated on ampicillin-agar medium. The bacterial colonies that develop are of two types, some originating from bacteria transformed by reformed vector DNA ($amp^R tet^R$) and some from those transformed by gene-vector DNA ($amp^R tet^S$), the *recombinant* colonies. To identify the recombinants, the colonies formed are *replica plated* on an agar-tetracycline medium. For this, a wooden block is carefully pressed on the total colony and removed with cells of each colony stuck to it in the same positon and arrangement as on the agar plate. This block is used as such for inoculation on an agar-tetracycline plate. Now only colonies of self ligated vector transformed bacteria ($amp^R tet^R$) appear, and colonies of the tetracycline sensitive recombinant bacteria ($amp^R tet^S$) are not formed. However, because of replica plating, the positions of the recombinant colonies can be located and identified in the original agar plate (Fig. 4.9). The bacteria from such recombinant colonies are scooped out, further cultured and pure gene-vector DNA in workable amounts isolated.

Recombinant selection by insertional inactivation of lacZ gene

The above method, involving two platings is quite tedious. An one step method can be done by using ampicillin resistance for screening of all transformed cells and simultaneously, β-galactosidase activity assay, coded by *lacZ* gene, for selection of only the recombinant cells. To appreciate the design, it is necessary to have an idea of the *lacZ* gene. Some *E. coli* strains, have a *lacZ* gene which codes for β-galactosidase enzyme. This enzyme breaks down lactose to glucose and galactose. In practice, this enzyme is assayed by using a lactose analog X-gal (5-bromo-4-chloro-3-indolyl-β-D-galactopyranoside) as a substrate, which yields a blue product. Colonies of bacteria having the *lacZ* gene, grown on agar turn blue, when X-gal and a galatosidase inducer as IPTG (isopropylthiogalactoside) are added to the medium.

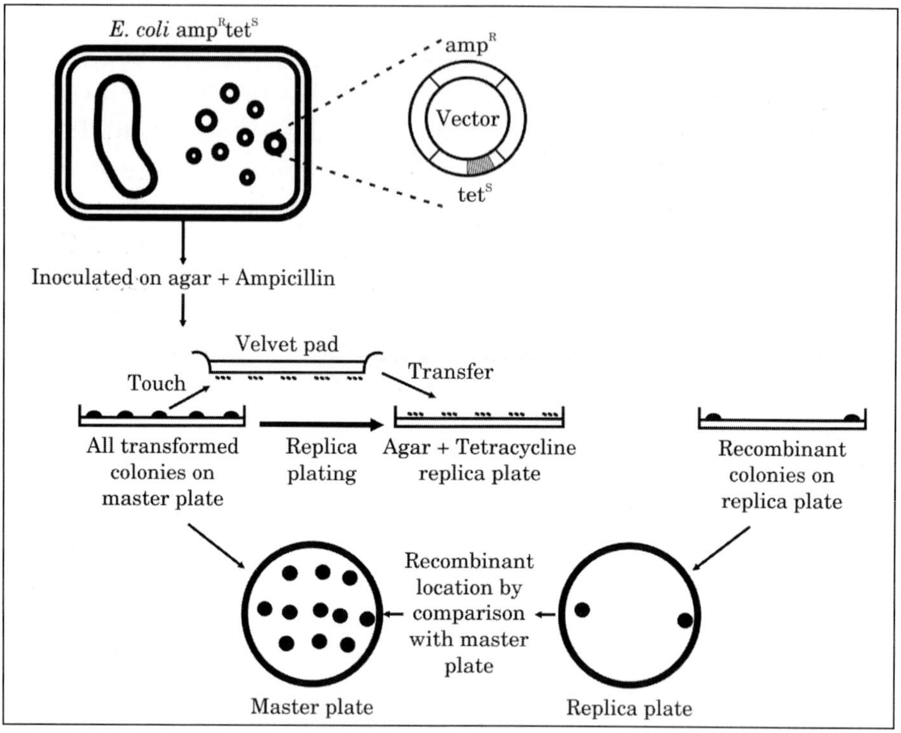

Fig. 4.9: Two step selection method of all transformed colonies first and then of recombinant colonies by replica plating

Difference between *lacZ* and *lacZ′* genes:

lacZ —*encodes*→ Alpha peptide + Beta peptide → β-galactosidase → Hydrolyzes lactose

lacZ′ → Alpha peptide only → Incomplete enzyme → Cannot hydrolyze lactose

The β–galactosidase enzyme consists of two sections, a smaller α-peptide at the N-terminal end and the larger β-peptide constituting the rest. While the *lacZ* gene synthesizes the whole enzyme, a mutant *lacZ′* gene codes for only the α–peptide (shown above). Some strains of *E. coli* besides being ampicillin sensitive, contain an incomplete *lacZ* gene, with the *lacZ′* segment missing ($amp^S gal^-$). These have inactive β-galactosidase, their colonies remaining white in agar media containing X-gal and IPTG. With such strains, a vector as the pUC8 vector having an ampicillin resistance gene and a *lacZ′* gene which can synthesize the missing part of the β-galactosidase enzyme of *amp*^{S}*gal*$^{-}$ *E. coli* strains, is used. On being transformed by such pUC8 plasmids, the cells become ampicillin resistant and have active galactosidase (*amp*^{R}*gal*$^{+}$). The colonies of such transformed cells turn blue on the addition of X-gal and IPTG.

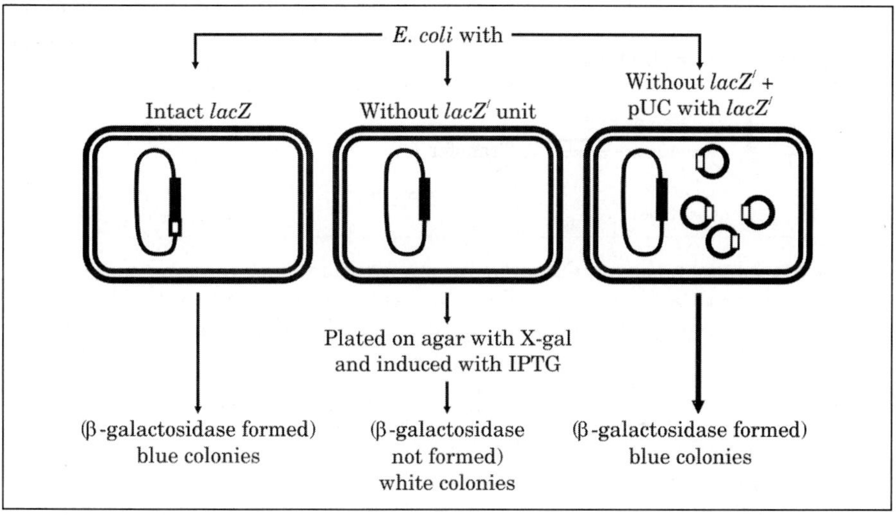

Fig. 4.10: Complementation between amp^{R}.gal^{-} *E. coli* and *lacZ*$^{-}$ pUC vector

For single step twin screening, the trick is to create a gap in the *lacZ'* site of pUC8 for insertion of the DNA to be cloned. Such insertion within the *lacZ'* gene causes *insertional inactivation* of this gene. The ligation reaction is carried out by mixing, ligase enzyme, pUC8 vector cut at *lacZ'* and the gene DNA molecules. The products will be mainly reformed vector, vector-vector and gene-vector recombinant molecules. An amp^{s}gal^{-} *E. coli* cell culture, incubated with such a mixture will undergo transformation by mainly reformed vector, vector-vector and gene-vector molecules. When plated on agar containing ampicillin, X-gal and IPTG, such incubated cells will form a mixture of colonies. Cells transformed by reformed pUC8 vector DNA (amp^{R}gal^{+}) will make galactosidase enzyme and turn blue. The recombinants however, will be white as in the gene-pUC8 recombinant DNA, the gene has caused insertional inactivation of the *lacZ'* gene causing amp^{R}gal^{-} transformation (Fig. 4.11). The

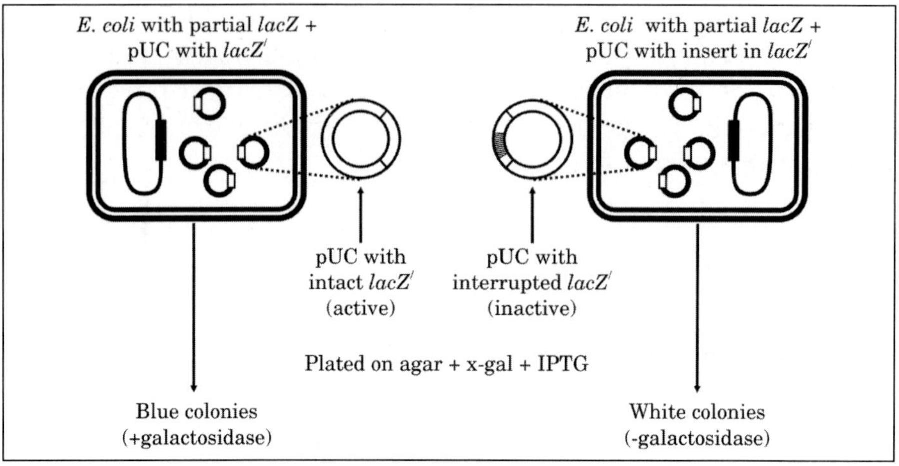

Fig. 4.11: Selection through insertional inactivation of pUC *lacZ⁻* gene by DNA insert Cloning of plasmid vector-gene recombinant in *E. coli* leads to massive amplification of the gene or DNA of interest.

recombinant bacteria from the white colonies are collected, further cultured and the recombinant DNA isolated.

Gene Cloning by Bacterial Transfection with Phage Vector-Gene Recombinant DNA

The parts of the lambda (λ) virus are shown in Fig. 4.12. Between the two cohesive ends are mainly, the segments for formation of head

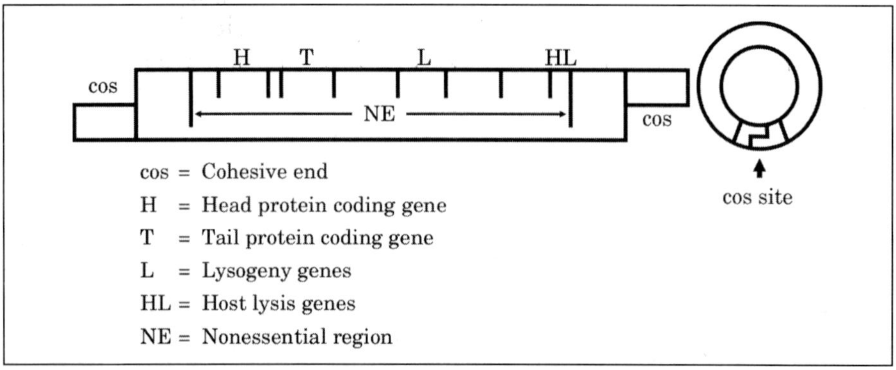

Fig. 4.12: (a) Lambda phage genome

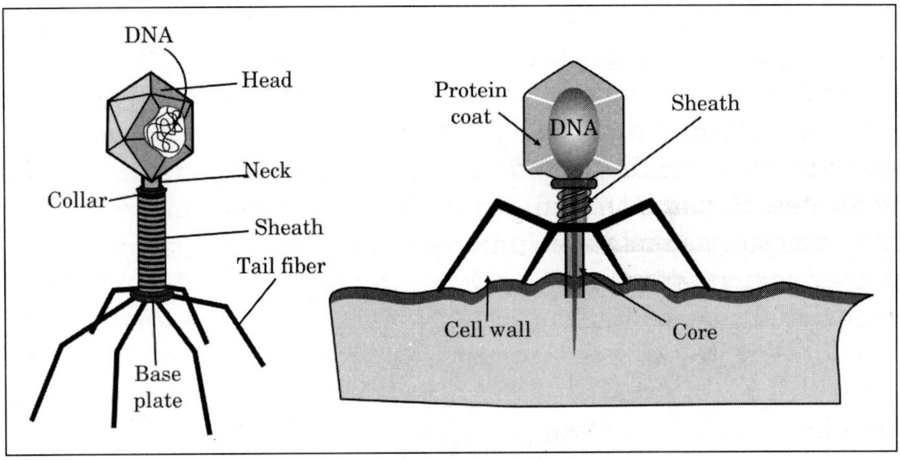

Fig. 4.12: (b) Lambda phage structure

and tail regions and for the processes of lysogeny and the lytic cycle. The λ DNA is about 49 kb long and has two cohesive, *cos*, ends. The ends are termed cohesive since, the two strands are complementary but unpaired, the strands *cohere* to each other.

Various types of vectors have been designed, by removing some parts of the virus, non essential for vector function. Some such vectors have been described in chapter 5. Recombinant DNA, is constructed by insertion of the gene or DNA of interest in the gap. The phage vector-gene recombinant DNA is introduced in suitable strains of *E. coli* for cloning and amplification.

Transfection

The recombinant phage DNA is incubated with competent *E .coli* cells and given a heat shock for uptake. This method is commonly used for introduction of a double stranded replicative form (RF) of an M13 cloning vector into *E. coli*. With λ phage, transfection is not efficient and the protocol of infection after *in vitro* packaging is used.

Transfection after in vitro packaging of lambda cloning vectors

At this point, a brief recap of the life cycle of this virus, may be useful. The λ virus after infection undergoes two phases, the *lysogeny cycle* and the *lytic* cycle. In the first, lysogeny pase, λ DNA integrates into the bacterial chromosome and exists as the prophage. The protein products of a *cI gene* in the λ DNA keep all the phage genes repressed. On deactivation of the *cI gene*, due to some stimuli, the phage enters the lytic cycle. The DNA dissociates from the main chromosome and replicates. Replication and synthesis of viral coat proteins start simultaneously. Replication initiates with formation of circular DNA, but, soon changes to a *rolling circle mechanism* with λ DNA chains connected end to end as concatamers. These chains are cut asymmetrically at regular intervals at sites destined to be the cohesive ends of released λ DNA. The λ DNA particles are packaged into heads and then the tails get added (Fig. 4.13).

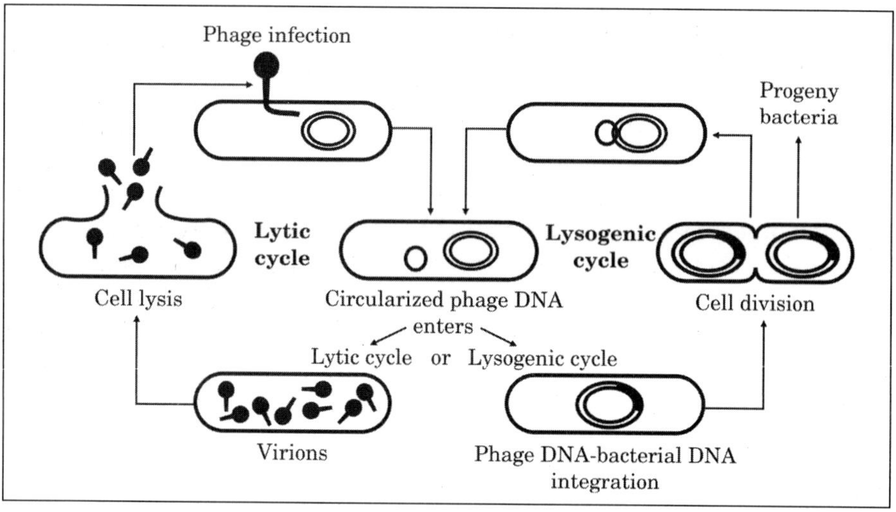

Fig. 4.13: Life cycle of λ phage

For *in vitro* packaging, the λ vector-DNA insert recombinant is mixed with capsid proteins in a test tube, which leads to the formation of packaged λ virus particles. But, how to get the capsid proteins? To obtain capsid proteins, *E. coli* cells are infected with lambda DNA mutated at the *cos* site. This lambda DNA cannot replicate but can synthesize all the capsid proteins. After requisite infection the capsid proteins are isolated from the bacterial culture and mixed in test tubes with the gene-lambda vector DNA for *in vitro* packaging and formation of whole viruses.

E. coli + Mutated unreplicable λ-phage —*Transfection*→ Capsid protein accumulation in bacterial cells —*Capsid protein isolated*→ Capsid protein

Capsid protein + Recombinant λ-Insert DNA —*In vitro incubation, packaging*→ *In vitro* Packaged lambda (λ) phage

An *E. coli* culture is infected with the packaged viruses and plated on agar medium. The uninfected bacteria form colonies and the infected ones, clear patches or *plaques*. The plaques are regions of bacteria infected and killed by the recombinant viruses. The plaques contain lysed bacteria and many phage particles. From the scraped material of plaques, phage DNA can be isolated. Considerable gene amplication is achieved by cloning with phage vectors also.

4.6. AMPLIFICATION OF RECOMBINANT DNA BY POLYMERASE CHAIN REACTION

This method was developed by Karry Mullis and has revolutionized recombinant DNA science.

Polymerase Chain Reaction (PCR)

PCR is an *in vitro* technique for making many copies of a DNA molecule in the test tube. This technique is for selective amplification of a delimited region of the DNA molecule. For synthesis of the selected sequence a special thermostable DNA polymerase I, *Taq polymerase*, is used. This enzyme being thermostable does not denature even at 90°C plus temperatures. It is obtained from a thermophilic bacterium *Thermus aquaticus,* which occurs in hotsprings. Besides this enzyme, two oligonucleotide primers, complementary to the $5'$ and $3'$ ends of the selected sequence are also needed. The primers delimit the sequence to be amplified. Thus, copies of specific DNA can be made by PCR.

For a PCR experiment, the sample DNA is mixed with *Taq polymerase*, primers and the four deoxynucleotides in a tube. The tube is heated to 94°C to denature the double stranded DNA into single strands and then cooled to 50 to 60°C to allow the primers to anneal to their respective complementary ends. The mixture is reheated to around 72°C, the optimum temperature of *Taq polymerase* to initiate DNA synthesis. This cycle of denaturation, annealing and DNA synthesis is repeated 25 to about 30 times for exponential amplication of the target DNA molecule. It must be pointed out that the primers have to be complementary to the $3'$ end of each strand, as, DNA synthesis proceeds in the $5' \rightarrow 3'$ direction. Thus, in each strand the primer attaches to the $3'$ end and the $5'$ end is free. Therefore, after the first cycle, long DNA strands, 'long products', with indeterminate $3'$ ends, extending beyond the base delimited by the primer in the other strand, are formed depending on where the synthesis stops. After 2nd cycle onwards, 'short products' delimited by the primers, that is the target sequence, are formed and accumulate. *Taq polymerase* lacks a major quality of normally occurring DNA polymerase, the ability of proof reading through its $3' \rightarrow 5'$ exonuclease activity. So, if a wrong deoxynucleotide gets inserted, it is not removed or proof read. This may slow down or even stop the PCR reactions. Therefore, besides this enzyme, another enzyme with proof reading capacity as, the enzyme *Tma* from *Thermotopa maritima* is also added to the PCR reaction mixture. The reaction proceeds at approximately 1000 bases per minute. If properly done, the ampilfication ratio, after 25 cycles can be 1 molecule: 4,194,304 molecules (Fig. 4.14).

For optimal PCR reaction, two factors are very important, the primer design and the annealing temperature. Conventionally, DNA molecule, is shown as having a 'top strand' with the base sequence read in the $5'$ to $3'$ direction from left to right and a complementary strand with the base sequence read again in the $5'$ to $3'$ direction, but from right to left. The primers have to be complementary to the $3'$ end of each strand. On the complementary strand template, the 'top strand' is synthesized, and the primer is referred to as the *forward*

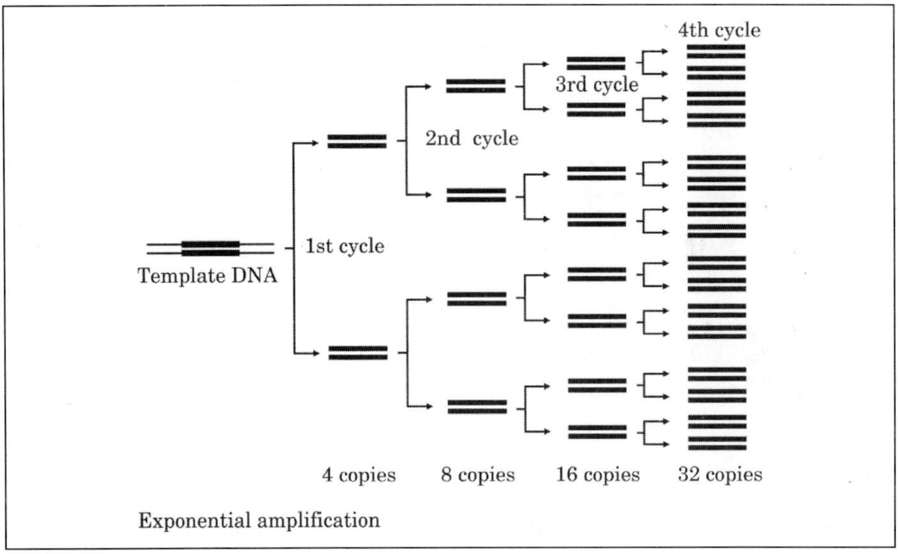

Fig. 4.14: Polymerase chain reaction (PCR)

primer and the one annealing to the top strand template is termed as the *reverse primer* as shown below.

Top strand	5′.....TGCATTAACGTAGGTTCCGGAA......3′
	Forward primer GGCCTT......5′
	5′.....TGCATT *Reverse primer*
Complementary strand	3′.....ACGTAATTGCATCCAAGGCCTT......5′

Care is taken to avoid complementarity between the forward and reverse primers to prevent primer-primer annealing. Further, the length of the primer oligonucleotide is adjusted to ensure that annealing occurs only once and at the 3′ border of the 'top' and the complementary strands and not on any site within the strand. Thus, primer length should be such that there are no attachment sites within the target DNA sequence besides the 3′ end borders of the two target strands. This is achieved by taking into account the bp length of the target DNA sequence. Suppose the target sequence has 10000 bases. If a primer, 5 bases long is used (5-mer) the possible attachment sites will occur once every 4^5= 4096 (4 bases^5) bases.

Thus, in a 10 kb target DNA there will be probability of the primer to attach to two sites, that is, besides the $3'$ border one more site. This will lead to the amplification of intermediate strands besides the main strand. Therefore, it will be preferable to use primers longer than 5-mers may be an 8-mer. Attachment sites for an 8-mer will occur every $4^8 = 65{,}536$ bases. So, in a 10 kb target DNA, the 8-mer primer will not have any chance of an attachment site besides the $3'$ border and therefore, only the target DNA will be amplified. In practice, primers of 17 to 30 bases are used. If the base sequence of the target DNA is known, the primers can easily be constructed. In case of an unknown sequence of a gene producing a known protein, use is made of known sequences of genes for homologous proteins.

The other important feature is the annealing temperature. It is usually kept about 2°C below the T_m or melting point of the primer-template hybrid. This can be experimentally determined or calculated by the formula (for short chains and the reaction being carried out in the presence of 50 mM monovalent cations):

$$T_m\ ^{o}C = (4 \times [G + C]\ ) + (2 \times [A+T])$$

In practice, PCR is carried out in programmable thermocyclers using piezoelectric heating-cooling system. The PCR mixture is kept in thin, 0.2 to 0.5 ml tubes and placed in metal blocks of the thermocycler. The cycler is programmed for automatic temperature transitions and the thin walls of the tubes help in rapid conduction of such heat changes. Water of the reaction mixture is prevented from evaporating away by either a layer of mineral oil which prevents evaporation or heating of tube caps which prevent any condensation of water on the lids. The PCR product/s is analysed by agarose gel electrophoresis.

Real Time PCR

There are various methods of quantifying the amount of PCR product. The most commonly used technique of quantitative PCR is *Real Time PCR*. In this technique the amount of the product formed can be estimated during the progress of the reaction. A common method is to use a *reporter probe* or *molecular beacon*. This probe is a short oligonucleotide, complementary to a region of the PCR product. The probe has complementary borders and is labeled with a fluorescent dye (*e.g.*, fluorescein) at one end and a quencher (*e.g.*, methyl red) at the other. Normally, the probe folds and the complementary borders anneal to bring the dye and the quencher molecules near each other. This prevents fluorescence. This probe is added to the PCR mixture and as products are formed the probe hybridizes with the complementary region of the product. This results in the probe unfolding and consequent removal of the quencher away from the dye, leading to fluorescence. Thus, the the product-probe hybrid emits a fluorescence signal, the intensity of the signal being proportional to the quantity of the product (Fig. 4.15).

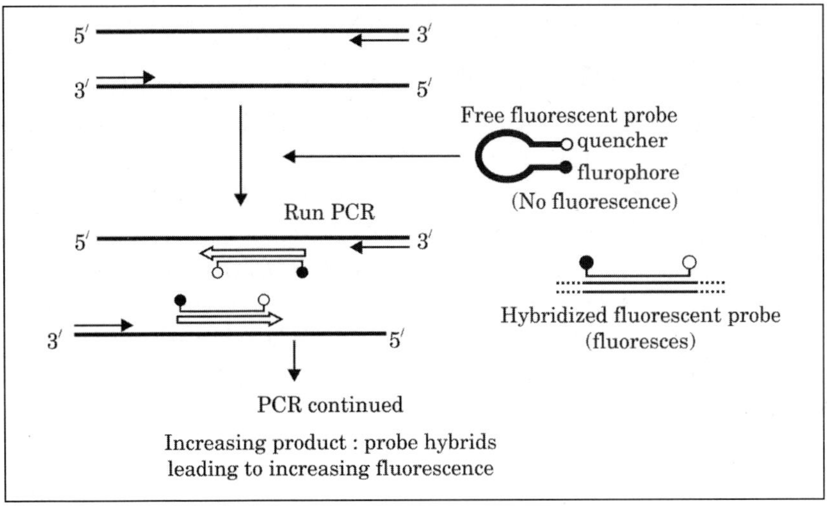

Fig. 4.15: Use of fluorescent probe molecular beacons for quantification of PCR

Quantification is done by comparison of the sample PCR with a control PCR of known amount of starting DNA. Usually the estimation is done by determining the number of cycles needed to reach a preset threshhold value. The higher the starting material, earlier the threshhold value is reached. As the PCR reaction and

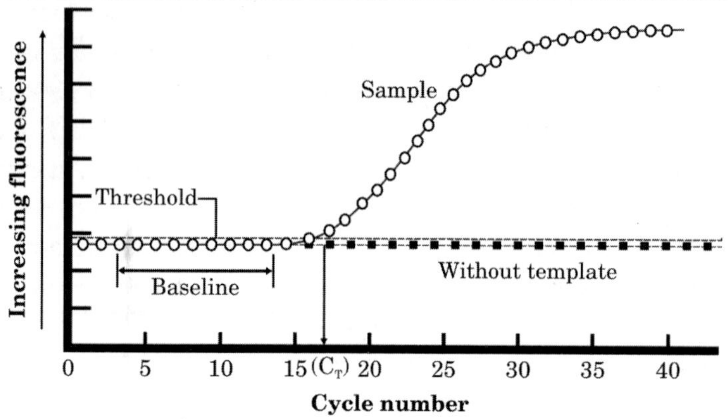

Fig. 4.16: Real time PCR with molecular beacon

reading of the fluorescence signal happen concurrently, the progress of the PCR is monitored in *real time*. In practice, the rise in fluorescence signal is plotted against cycle number (Fig. 4.16).

The plot reveals that there are three stages, first a slow accumulation, then an exponential rise and finally a plateau. As mentioned, a threshold value is set above the baseline. The parameter C_T, the cycle number at which the fluorescence rises above the threshold, is determined. The amount of template DNA at the start (T_0) and at a cycle number n (T_n) is calculated from the formula:

$T_n = T_0 (E)^n$, where E is the efficiency of amplication. The maximum theoretical E value can be 2 assuming that all templates are duplicated always. Actually it is lesser than 2, like 1.5 to 1.7.

Cloning PCR Products

Sometimes it may be necessary to clone a PCR product. There are two methods. In the first method advantage is taken of the fact that PCR reactions usually add dA at the 3′ end leaving 3′ A overhangs. Such products can be ligated to commercially available vectors with 3′ T overhangs and further processed for cloning. In the second method a primer with a short oligonucleotide containing a restriction sequence is added to the 5′ end of each primer. This leads to addition of a tail with restriction site to each strand of the PCR products. Such products can be ligated to any vector of choice and the recombinant DNA cloned.

4.7. DNA SEQUENCING

The linear sequence of monomers in a biopolymer, from one defined end to the other is referred to as its *sequence*. Its determination is necessary for understanding correlations between the polymer's stuctural aspects and function. This is true also for DNA, the chemical basis of heredity. DNA sequence is the linear sequence of bases from the 5′ end to the 3′ end. It must be noted that a DNA sequence for a gene means the sequence of the coding strand, which shows the sequence of codons from 5′ → 3′ end, a sequence which is colinear with the amino acid sequence of the protein it codes. Some popular sequencing methods are described briefly in the following sections.

Chain Termination Method

This method was developed by Sanger and his colleagues. Normally in cells, for replication of a DNA strand in the 5′ → 3′ direction, a primer attaches to the 3′ end of the template strand. The deoxy nucleotides from the cellular medium attach to their complementary bases from the 3′-OH end of the primer onwards and are linked through phosphodiester bonds by DNA polymerase to make the complete strand. The Sanger method is based on termination of the synthesis at regular intervals by dideoxyribonucleotides (ddNTP) which are analogues of the deoxyribonucleotide (dNTP) bases as shown below in a simplified form.

As shown in the figure the sample strand is annealed to a primer, and put in a reaction mixture containing dA, dT, dG, dC and a special DNA polymerase *sequenase*. Four such sets are made and the reaction started. From each set aliquots are taken at different intervals and a particular ddNTP is added. This addition of ddNTP stops or

Chain termination sequencing:

5′------------------------------3′ (hypothetical sample) + dATP, dTTP, dGTP, dCTP + ddATP, ddTTP, dGTP, ddCTP, (in small amounts and labelled with fluorescence labels emitting different fluorescence colours) + DNA polymerase (sequenase)

Sequences obtained by Gel electrophoresis and scanning are:

Sample DNA as template

5′---------------------------3′
A-5′
G-A-5′
A-G-A-5′
G-A-G-A-5′
C-G-A-G-A-5′
G-C-G-A-G-A-5′
A-G-C-G-A-G-A-5′
C-A-G-C-G-A-G-A-5′
3′-G-A-C-A-G- C-G-A-G-A-5′

Full sequence obtained is:
3-G-A-C-A-G-C-G-A-G-A-5′
Therefore, the sequence of the sample template strand is:
5′-C-T-G-T-C-G-C-T-C-T-5′

terminates synthesis as ddNTP is not recognized by the polymerase enzyme. Thus, over time, one gets four sets of a series of aliquots with the synthesis stopped by ddT or ddA or ddG or ddC. The aliquots of each set are pooled and subjected to polyacrylamide gel electrophoresis in capillaries. The ddNTPs are tagged with fluorescent markers, each with a different marker and the electrophoretic separation of the bands recorded by a fluorescence detector. From the lengths of the chains terminated by ddNTP, the sequence can be read.

The method requires single stranded DNA. Such single stranded DNA is obtained by cloning the DNA with a phagemid vector. The other option is conducting the sequencing reaction with a thermostable DNA polymerase in a thermocycler.

In case of cloned DNA, a universal primer which may anneal to a short stretch of the vector DNA, adjacent to the DNA of interest may be used. In case, of DNA of known protein product, besides the starting primer, some short internal primers can be synthesized and used to hasten and complete the sequencing process.

Pyrosequencing

The basic strategy of this protocol is, coupling of nucleotide addition to a DNA strand, by phosphodiester bond formation, with chemiluminiscence.

Pyrosequencing basic concept: Nucleotide addition coupled to Chemiluminiscence

primer

-------A------------------ + TTP —DNA Polymerase→ -----T / -----A------------- + PPi —ATP-Sulfurylase→ ATP —Luciferase→ **light emission**

DNA sample; **T added**

In this method, DNA biosynthesis on the sample DNA strand, with a primer annealed, is carried out in the presence of a suitable DNA polymerase. The mixture contains also sulfurylase and nucleotidase. Now, each deoxynucleotide is added one after another. One of the dNTP gets bonded to the preceding dNTP leading to the release of pyrophosphate. The immediate conversion of the pyrophosphate leads to a chemiluminiscence which can be detected. The three other unincorporated dNTPs are degraded by nucleotidase. So, chemiluminiscence indicates the incorporation of a nucleotide. By serial additions and recording of the nucleotide added the sequence can be read as shown below.

Pyrosequencing:

1st Reaction

primer
5′-------
3′------------------5′ + DNA polymerase
Sample DNA

+ dTTP → No light emission
+ dGTP → **Light emission**
+ dCTP → No light emission
+ dATP → No light emission

→ 5′-----------G
3′-----------------------5′
product 1

2nd Reaction

5′-------**G**
3′------------------5′ + DNA polymerase
Product 1

+ dTTP → **Light emission**
+ dGTP → No light emission
+ dCTP → No light emission
+ dATP → No light emission

→ 5′-----------**GT**
3′-----------------------5′

Suppose the nth reaction product is 5′-------GTCAAGCTCG-3′ New strand
3′-------CAGTTCGAGC-5′ Template strand

Then the sequence of the sample DNA strand in 5′ → 3′ direction is: **5′CGAGCTTGAC-3′**

(unutilized nucleotides, of each step, are degraded by a nucleotidase mix. apyrase)

Pyrosequencing can be done in both liquid and solid phases. This method has been largely adopted for automated sequencing of DNA.

4.8. FRACTIONATION AND DETECTION OF MACROMOLECULES

For detailed studies of individual macromolecules like DNA, RNA or protein, total isolation is followed by fractionation. As these are charged molecules, the bulk cellular extract of any of these can be separated by electrophoresis. The commonest electrophoretic procedure is gel electrophoresis. The separated bands are then detected by staining. For identification of specific molecules, techniques based on hybridization with probes as complementary oligonucleotide strands in case of nucleic acids and antibodies for proteins are commonly used.

Gel Electrophoresis

Two types of gel matrices are used agarose and polyacrylamide:

Agarose gel electrophoresis

The equipment consists of a glass or plastic plate resting with their edges on tanks containing suitable buffer (Fig. 4.3). An electrode dips in each tank and the two electrodes, an anode and a cathode, are connected to a DC power supply. Generally a 0.5% agarose gel is prepared. The warm liquid gel is poured on the plate and allowed to set. The thickness of the gel is maintained at around 5 mm. While pouring the gel a specially designed comb is placed with teeth down accross the plate near the anode end. After setting of the gel, the comb is carefully removed to expose the sample wells formed. The gel is connected on both sides to the buffer in the tank with filter papers. The samples in a suitable medium are pipetted into the wells, a tracking dye, as, bromophenol blue is added and the electrophoretic run started. The movement of the run can be tracked by the movement of the dye front and the electrophoretic run is stopped when the dye frontline reaches the other end of the plate. Separated bands can be seen by staining with specific dyes. For instance DNA bands are visualized by bathing the plate in an ethidium bromide solution and exposure to UV light. Similarly, protein bands can be detected with comassie blue.

Polyacrylamide gel electrophoresis

The setup is as above except that instead of agarose, polyacrylamide is used. A detergent, SDS (sodium dodecyl sulphate) is added in case of protein fractionation.

Detection of Specific Bands by Blotting

The techniques described below for detection of specific DNA, RNA and protein molecules are based on complementarity or specific site recognition dependent hybridization or *blotting*. The first such technique was designed for DNA by Professor F.W. Southern and is therefore referred to as *Southern blotting*. Similar protocols were developed by other scientists for RNA and proteins. Interestingly it seems, in honour of Professor Southern, the naming of these latter methods have been based on compass directions than the discoverers! The protocol for RNA is called *Northern blotting* and that for protein as *Western* blotting.

Southern blotting/Southern hybridization

After the DNA fragments of a specific restriction enzyme digest are separated on a gel, they have to be searched for a specific molecule, like a single stranded gene segment out of several bands. An approach can be to bathe the gel with a solution of labeled single stranded probe DNA which is complementary to the gene DNA. Autoradiography of the gel can show the hybrid band. However, such gel assays were of very low resolution and fuzzy because of background hybridization. Therefore, Professor Southern devised the technique of transferring the DNA bands from the gel to a nylon or nitrocellulose membrane. For this, after the electrophoretic run, a requisite membrane is placed on the gel and several layers of filter papers are overlaid on the membrane. The paper heap is kept pressed by

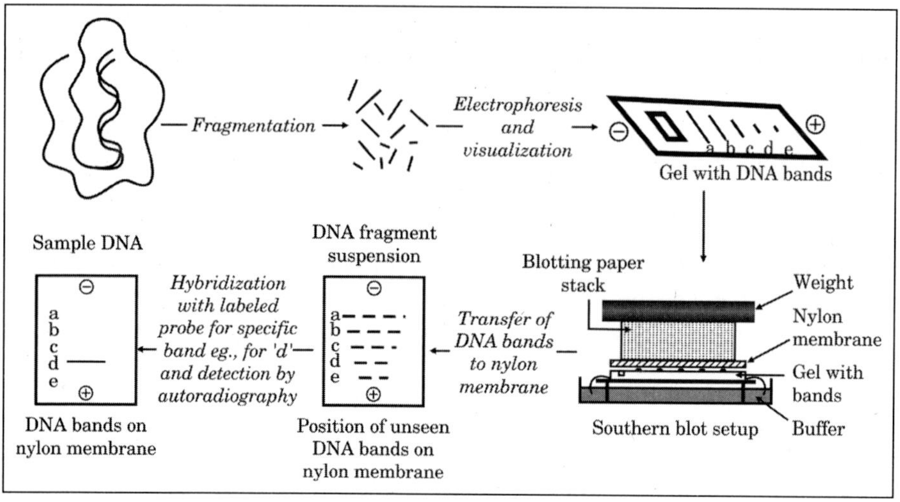

Fig. 4.17: Southern blotting

a weight (Fig. 4.17). The buffer rises from the gel and moves upward by capillary action towards the filter papers and in the process transfers the DNA bands to the membrane. The gene or the DNA of interest is visualized by autoradiography in case of radio isotope or a suitable assay for non radioactive labels.

Northern blotting or northern hybridization

This is the name of the technique for detecting RNA specially mRNA by a similar setup and protocol as above. The probes used are synthetic labeled complementary RNA.

Western blot or western hybridization

The basic outlay is same as Southern hybridization. Protein bands are transferred to nitrocellulose or nylon membrane and identified by particular protein specific antibody. Prior to the analysis, antibody against the protein antigen of interest is raised in experimental animals as rabbits. Rabbits are injected with the specific protein and the antibody allowed to develop. After due period blood samples are drawn out from such rabbits without killing them, the antibody collected and purified. The antibody is applied on the protein containing gel and binds to the protein of interest and can be identified. In practice the identification is done by using labeled secondary antibodies which bind to the primary antibody and further heighten the signal for autoradiography or other form of detection.

4.9. TRACER TECHNIQUES

For many studies on metabolism and dynamic processes, it becomes necessary to follow or trace the stepwise changes in an element or a compound. For such studies, since many years, radio or heavy isotopes of elements have been of great use. Isotopes are forms of an element with same number of protons, but, different numbers of neutrons, that is, of same atomic number, but, different atomic mass. Radio isotopes are more commonly used. A metabolite is *labeled* with a radio isotope by replacing one of its atoms by a radio isotope and then the fate of the radioactive metabolite can be *traced* by various detection methods. The technique is termed as *tracer technique.* Some examples of radioisotopes ($X^{/}$) are shown with the normal version (X) in parentheses alongside as, $\mathbf{X}^{/}$ (**X**) with atomic mass as superscript on the left and atomic number as subscript on the right (the atomic numbers shown here as subscript on the right is for convenience and is not as per convention): $^{3}H_1$ / ($^{1}H_1$), $^{14}C_6$ / ($^{12}\mathbf{C}_6$), $^{32}P_{15}$ / ($^{30}P_{15}$), $^{35}S_{16}$ / ($^{32}S_{16}$). The radio isotopes have an unstable combination of neutrons and protons and gradually disintegrate. During disintegration various radiations are emitted, as, α-particles (alpha, 2 protons and 2 neutrons), β-particles (beta, electrons) and γ-radiations (gamma, electro magnetic radiations). The common radioisotopes used are beta emitters because of the ease of their detection. As, radioisotopes degenerate, information of their life span, estimated by their half life, is necessary for experimental purposes. The isotopes mentioned above, have sufficiently long half lives for practical applications (^{32}P-14 days, ^{35}S-87 days, ^{3}H -12 years, ^{14}C-5700 years). For quantitave analysis, detection is done in *scintillation counters.* In the counter, beta rays from the radioactive sample kept in a cuvette, pass into a solution of *scintillant* molecules activating them to emit light, the light emitted falls on a photocell and is converted to a proportional amount of current which is, graphically recorded. Such quantification is possible because, from the primary intensity of beta radiation to the extent of final electrical pulse, all are proportional to the amount of starting radioactive material.

Visualization of the location of labeled compounds in an organ or tissue section and labeled bands of macromolecules in electrophoretic plates, is done by *Autoradiography.* In brief, the sample slides or matrices as, nitrocellulose filters with labeled sample bands, are pressed against photographic plates in light proof casings. After due period, development of the plates reveal the labeled locations or bands as dark regions in a light background. Such detection is possible because photographic plates are sensitive to the beta rays.

4.10. DNA MICROARRAY TECHNIQUE

The above methods have enabled great advances in DNA science. However, investigations on large complex genomes, required more rapid protocols. DNA microarray technology has been a very helpful innovation in this direction, for rapid analysis of whole genomes.

In this technique, known DNA sequences as, cDNAs or smaller oligonucleotide portions of cDNAs, the ESTs (expressed sequence tags) are immobilized on a glass slide or membrane filter in lines or array. Microarrays are also made by fixing or 'printing' many DNA sequences on substrates as, glass slides through robotics to construct DNA chips. Such microarrays contain thousands of sequences of a gene in a spot of about 1000 nm diameter (Fig. 4.18).

DNA microarray analysis uses the principle of hybridization between complementary DNA strands. The known DNA sequences immobilized on the microarray are the probe molecules and the DNAs of the sample are the 'target' molecules to be studied. For such analyses, first of all microarrays have to be constructed or the complex ones purchased. RNA is extracted from the sample and converted to cDNA by reverse transcriptase. The cDNA mix is treated with a fluorophore to label each cDNA present. The labeled cDNA solution

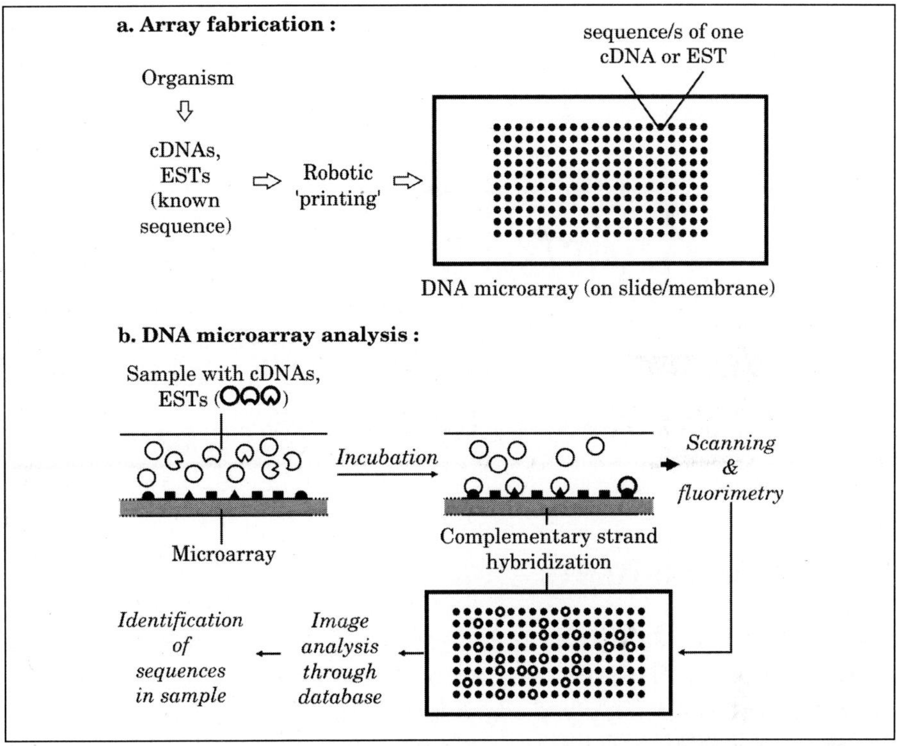

Fig. 4.18: DNA microarray and microarray analyses

is incubated with the microarray for required period, washed, dried, subjected to fluorescence scanning and the image processed for data acquisition. The target sequences complementary to the fixed probes are revealed through fluorescent spots of the hybrids formed. When comparisons have to be made, *e.g.,* between healthy and diseased tissues, the cDNA pool of each is labeled by different fluoroprobes emitting different colours.

Various techniques, specially innovations in techniques, form the foundation of experimental science. Innovations in technology emerging from the spirit of enquiry lead to ever deeper understanding of nature. The above account, although not complete, will give some idea of the efforts that have been made in understanding molecular biology and the recombinant DNA science.

5

Vectors for Prokaryotic and Eukaryotic Cells

SUMMARY OF CONTENTS

Cloning Vectors for Prokaryotes- plasmid vectors, phage vectors, cosmids, Bacterial Artificial Chromosome (BAC); Cloning Vectors for Eukaryotes - BAC, yeast episomal plasmid (Yep), yeast artificial chromosome (YAC); Expression Vectors for *E. coli*, yeast, plant cell and insect cells (baculovirus).

Large scale cloning of both coding and non-coding DNA is necessary for detailed correlative studies on structure and function of various stretches of DNA. The insights from such studies form also the foundation for beneficial applications. As is evident from the previous chapters, *cloning* or amplification of a DNA fragment into numerous copies is necessary for obtaining DNA in workable amounts. This is done by making a DNA-vector recombinant DNA and introducing it in a suitable cell where it can replicate and multiply. Thus, a clone is a colony of bacteria. The clone may be of a gene, a regulatory element, intron, or some non-coding DNA.

The most commonly used prokaryotic cell for cloning is *Escherichia coli.* However, for many studies on larger inserts and also for commercial purposes mainly yeast cells have been used. The following sections describe some of the most commonly used vectors. Vectors based on their use are of two types, the *cloning vectors* and *expression vectors.* The cloning vectors are for DNA amplification as clones and the genes inserted cannot encode proteins. This is because the promoter necessary for gene expression is not inserted in the vector. In expression vectors, on the other hand, the gene inserted is accompanied by its promoter to ensure its expression.

The developments in vector designing have been motivated by various requirements. The main objective has been to increase the carrying capacity of vectors. Thus, amongst cloning vectors, while the plasmid vectors carry DNA inserts 5 to 10 kb long, the BAC (bacterial artificial chromosome) vectors can accommodate DNA inserts of 100 to 300 kb length. The other major requirement has been the host range, designing vectors for use in different organisms. Vectors have been developed for functioning in various plants and animals. Such efforts have led to construction of many expression vectors for production of important biomaterials.

5.1. CLONING VECTORS FOR PROKARYOTES (MAINLY *E. COLI*)

Normally, for genomic studies the clones obtained from the genome of a species are stored as a *genomic library*. Thus, the clones in a genomic library are for experimenters, like books in a library. However, unlike books, the clones are untitled and have to be identified or *screened*. There are several types of vectors used for cloning DNAs of interest in bacteria. The main bacterial host is *E. coli*.

Plasmid Vectors

Plasmids are extrachromosomal circular DNA of some bacteria. They replicate and multiply in the bacteria and may confer some traits to the host like antibiotic resistance, pigment production, nitrogen fixation, tolerance to heavy metals and other traits of value to the host. The trait conferred to the host bacterium can get transferred to other bacteria by movement of plasmids to other bacteria through conjugation (Fig. 5.1).

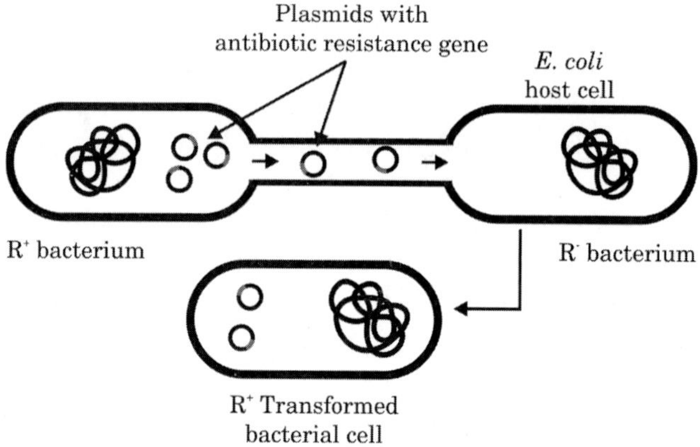

Fig. 5.1: Bacterial transformation by transfer of plasmid borne antibiotic resistance gene from another bacterium

A plasmid sequence, therefore, has a region governing replication referred to as origin of replication or ori, one or more antibiotic resistance genes, some restriction enzyme recognition sites and other sections. Usually there are segments in the sequence, removal or replacement of which does not affect the survival or replication of the plasmid. In such a section, a DNA of interest can be inserted. Plasmids can have their replication under *stringent* control of the main chromosome or *independent* of it. In the former case, the plasmids multiply along with the main chromosome and in the latter independently. Plasmids may have narrow or broad host range. For purposes of vector construction, plasmids with independent replication, some marker genes as those of antibiotic resistance and required host range are selected. Further, a *multiple cloning site* (MCS) is inserted in a suitable location to allow convenient insertion of a gene or any DNA sequence of interest (Fig. 5.2).

Sometimes, two Ori segments are inserted, for instance, one for *E. coli* and the other for yeast to ensure that the plasmid can replicate in both the hosts. Such vectors, the *shuttle vectors* are very useful.

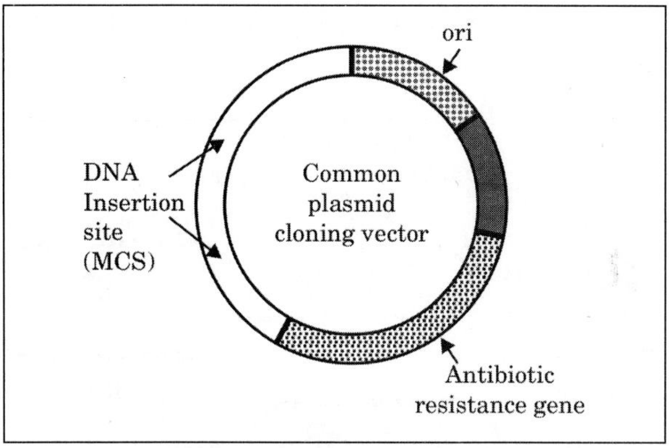

Fig. 5.2: Essential components of a plasmid vector

The criteria for suitability of a vector include small size, efficiency in inducing transformation, distinct markers for selection of transformants and recombinants and carrying capacity, that is the maximum size of the DNA fragment that can be inserted.

pBR series vectors

The pBR series of plasmids were developed in early 1970s. The earliest vectors constructed were pBR322, pBR328 and others. pBR322 is 4.36 kb in size and derived its elements from three natural plasmids. The ampicillin and tetracycline marker genes are from plasmid R1(pRS2124) and pSC101 respectively and the replication sequence from pMB1. This vector (Fig. 5.3) has two selection markers, amp^R and tet^R genes, each with several restriction sites including 6 unique sites in the former and 11 unique sites in the latter. At either site, a DNA fragment can be inserted and identified through a selection method based on insertional inactivation. DNA fragments upto 6 kb in length can be inserted.

The rules for naming of vectors can be understood from the example of pBR322. In the name assigned to this vector, 'p' stands for plasmid. The letters in upper case are mainly for the inventors or place of origin. Thus, in pBR, B and R are for Bolivar and Rodriguez, the designers. The number 322 is a specific identification

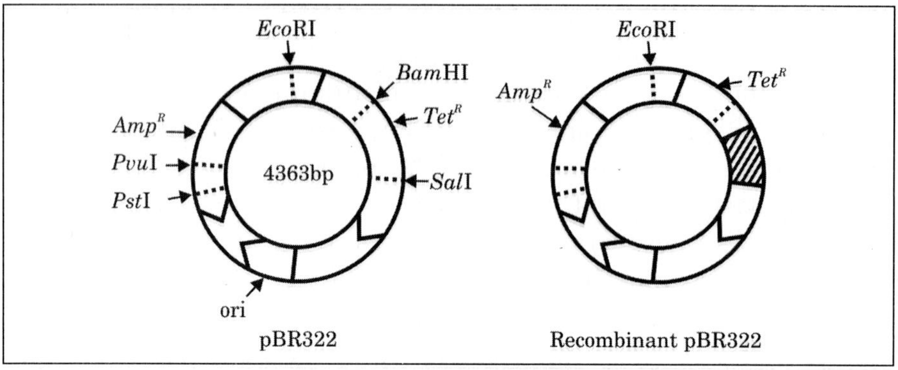

Fig. 5.3: Generalized structure of a pBR322 vector and recombinant pBR322 (Amp^R = ampicillin resistance gene, Tet^R = tetracycline resistance gene, ori = origin of replication, other labels are of restriction enzymes)

number as, there are other simlar derivatives as pBR327, 328. The variant pBR327 has been derived by deletion of a 1089 bp section and has the advantage of being *nonconjugative*. This prevents the transformed *E. coli* to pass on the gene to another *E. coli* cell, thus preventing unwanted gene transfer or *gene contamination.* Such designing ensures the much needed *gene containment*. The pBR322 design laid the foundation for the development of better vectors in future. All pBR based recombinants have to be identified by a two step selection method, by plating first in ampicillin agar medium and then replica plating on a tetracycline containing medium (Fig. 5.4).

pUC series vectors

In 1980s came the better designed pUC series of vectors, pUC8, pUC9, pUC18 and others. These vectors, derived partly from pBR322, are smaller being about 2.75 kb in size, have a *lacZ′* marker besides the ampicillin resistance marker and specially a *multiple cloning site* in the *lacZ′* segment for more versatile gene insertion. The Ori sequence allows for high copy yield (Fig. 5.5).

The basis for selection is α-complementation as mentioned in chapter 4 also. The *lacZ′* gene encodes the 146 amino acid long N-

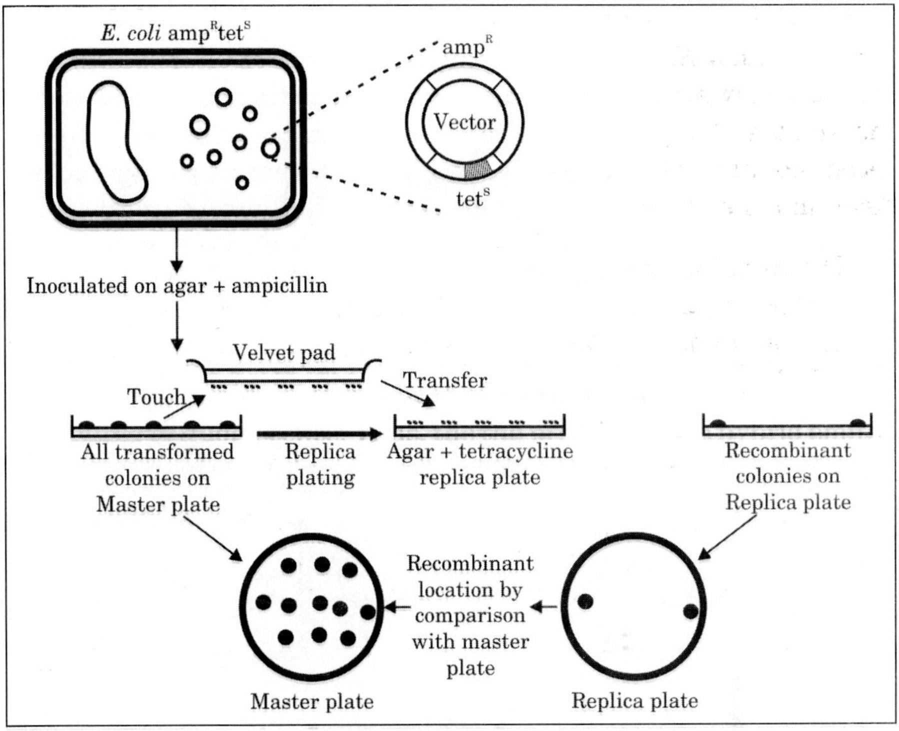

Fig. 5.4: Two step identification of recombinant bacteria by replica plating

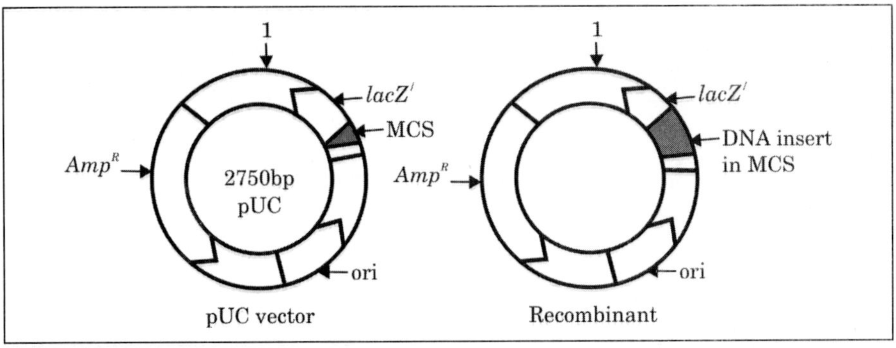

Fig. 5.5: Generalized structure of pUC series vector and recombinant

terminal α part of the β-galactosidase enzyme. Similarly, a mutant of *E. coli* (*e.g.*, *lac*Z'M15) encodes the remaining part of this enzyme from the C-terminal end. When the pUC vector with *lacZ'* enters an *E. coli lacZ'*M15 cell, there is complementation between the two partial *lacZ* genes leading to synthesis of the complete β-galactosidase enzyme. Such *E. coli* colonies turn blue in X-gal containing agar medium. However, in pUC vectors, DNA insertion in *lac Z'* segment inactivates the gene. Mutant *E. coli* cells transformed by such recombinant pUC with inactive lacZ', cannot synthesize β-galactosidase and the colonies remain white on X-gal-agar medium.

The combination of amp^R and *lacZ'* markers allows for a single step selection, all transformants being selected by ampicillin and the recombinants by blue/white screening on the same agar plate (Fig. 5.6).

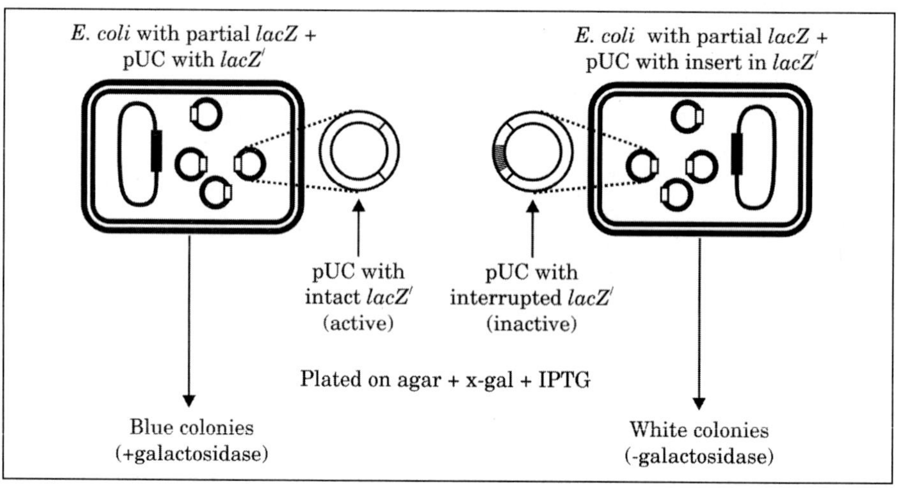

Fig. 5.6: One step identification of recombinant bacteria by insertional inactivation of *lacZ'* gene in pUC vector by DNA insert

Phage Vectors

The plasmid vectors have limitations regarding size and form of the DNA insert they can carry. Thus, plasmid vectors can accommodate DNA inserts of maximum about 8 kb length. Further, the amplification is mainly in the double stranded replicative form (RF). Therefore, it was necessary to develop firstly vectors capable of carrying larger inserts. As, in many studies like sequencing, single stranded DNA is needed, vectors were also designed for amplification of DNA in single stranded form. To serve these purposes phage based vectors were found to be quite useful. There are various vectors based on specific potentials of different bacteriophages. The main phage vectors have been derived from the filamentous phage M13 for mainly single stranded DNA amplification and lambda (λ) phage for larger DNA inserts.

M13 based vectors for DNA amplification in single stranded form

Filamentous viruses transfecting bacteria, as the M13 bacteriophage have single stranded (ss) DNA and can amplify DNA inserts in the single stranded form. Such ss DNAs are much needed for several studies like sequencing, making of probes and oligonucleotide mutagenesis.

The circular M13 ss DNA is packed with a row of genes which are indispensable. This segment is connected by an intergenic strand of about 500 bases length and which contains the ori sequence also. After transfection of enteric bacteria, the ssDNA is converted to double stranded (ds) replicative form (RF). This dsRF DNA replicates for some time and then a protein encoded by itself binds to one of the strands. This induces a nick and stops its replication. The other free strand continues to replicate by a repetitive *rolling circle* mechanism. Splitting of the new strand at requisite points and circularization leads to formation of daughter circular ssDNAs. These DNAs acquire the coat proteins to form the progeny virions. The progeny viruses are released through the host bacterial membrane. A special feature is that completion of life cycle of the transfecting M13 does not kill the host bacterium. The transfected host continues to divide, at a slower pace, and releases many phage particles at each generation thus providing a good system for ssDNA amplification (Fig. 5.7).

Several M13 vectors have been created by insertion of *LacZ'* gene and its modification. The M13mp2 has a *LacZ'* gene at the starting point of which a GAATTC *Eco*RI site has been created by *in vitro* mutagenesis of a naturally occurring GGAATTC sequence (Fig. 5.8). A DNA strand can be inserted at the restriction site. Several vectors as M13mp7 have been constructed by further inserting a *polylinker*

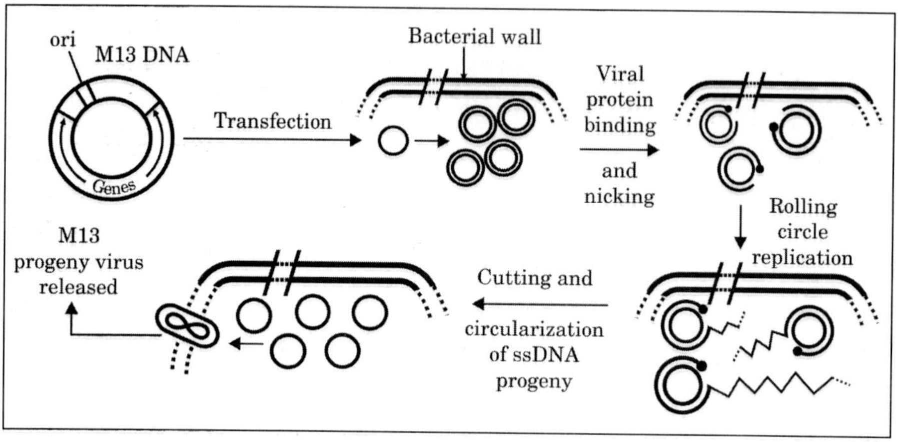

Fig. 5.7: M13 phage life cycle

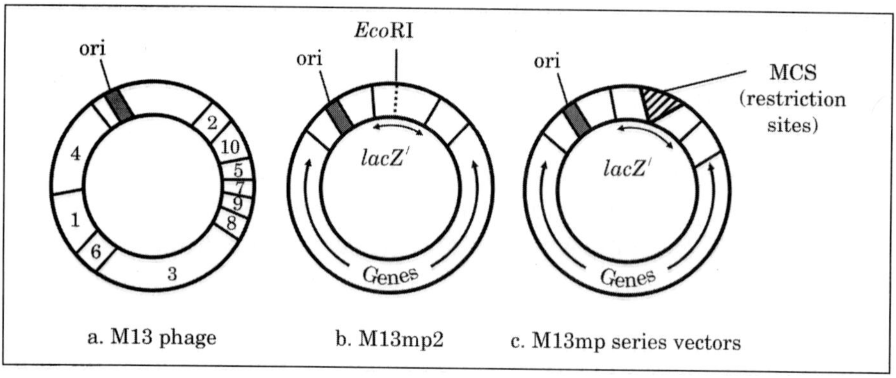

Fig. 5.8: (a) M13 phage genome, (b) M13mp2 with *lacZ'* containing an EcoRI site, (c) Generalized M13 mp series vectors with MCS inserted in *lacZ'* EcoR1 site

in the *lacZ'* sequence without effectively disrupting its coding activity. It has four restriction sites (*Eco*RI, *Bam*HI, *Sal*I and *Pst*I). The transformation events can be detected by functional β-galactosidase activity using X-gal.

Although M13 based vectors release single stranded DNA, the size of foreign DNA to be cloned is limited to little more than 2 kb. To overcome this constraint, some hybrid vectors, the *phagemids*, have been constructed by fusion of a pUC vector with a M13 fragment containing the signal sequence for conversion of ds DNA to ss DNA and the phage ori.

λ phage based vectors

The phage is 49 kb long. Fortunately the mid region of the core DNA, of about 15 kb is not necessary for replication in bacteria. For vector construction, this section is used for insertion of the DNA to be cloned. The section deleted has sequences responsible for integration of the phage DNA to *E. coli* chromosome as also subsequent excision, elements which enable the phage to enter into lysogeny. Lysogeny is unnecessary for cloning purposes and the deletion makes the phage *nonlysogenic* allowing the recombinant phage to enter directly the lytic cycle. The natural phage has many restriction sites distributed all over, which make it inconvenient as a vector. So, natural selection, by repeated culture is done to obtain strains with none or few restriction sites. The selected λ DNA molecules without restriction sites and with necessary modifications as deletion of nonessential region are used to make insertion and replacement vectors (Fig. 5.9).

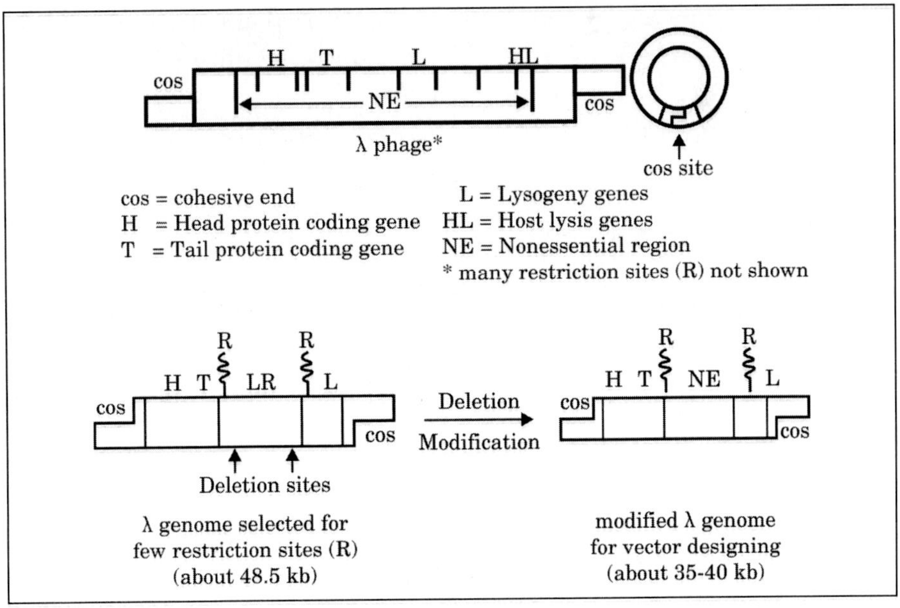

Fig. 5.9: λ genome modification for vector designing (H=head, T=tail, LR= lysogeny and regulation, L=lysis, NE=nonessential residue)

Insertion vectors

For constructing insertion vectors, the modified λ genome has one restriction site *e.g.* for the enzyme *Eco*RI. In one common vector, λgt10, the sample DNA is inserted in the *Eco*RI site in the *cI* gene and detected by formation of clear rather than turbid plaques by recombinants. The other λ ZAPII vector, has a polylinker, with several restriction sites, inserted in the *lacZ'* gene. The DNA of interest is inserted in the polylinker and recombinants tested by X-gal (Fig. 5.10).

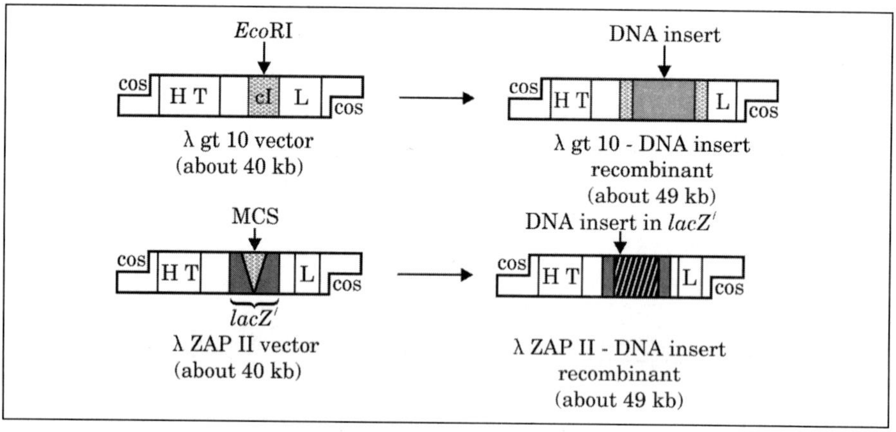

Fig. 5.10: λ Phage based insertion vectors λgt10 (*Eco*R1 site in *cI* gene) and λZAPII (polylinker in *lacZ'* site) and recombinants

The carrying capacity of insertion vectors is about 5 to 10 kb long DNA inserts.

Replacement vectors

These vectors are used for cloning of larger upto about 20 kb long DNA. Construction of replacement vectors involves use of a modified λ genome with two restriction sites of the same enzyme. Advantage is taken of the restriction sites flanking a nonessential *stuffer* segment of the λ DNA. Using any one of the restriction enzymes, the segment is removed leaving sticky ends. The DNA of interest with similar sticky ends generated, replaces the segment, anneals and is ligated. One commonly used vector is λ EMBL4 (Fig. 5.11).

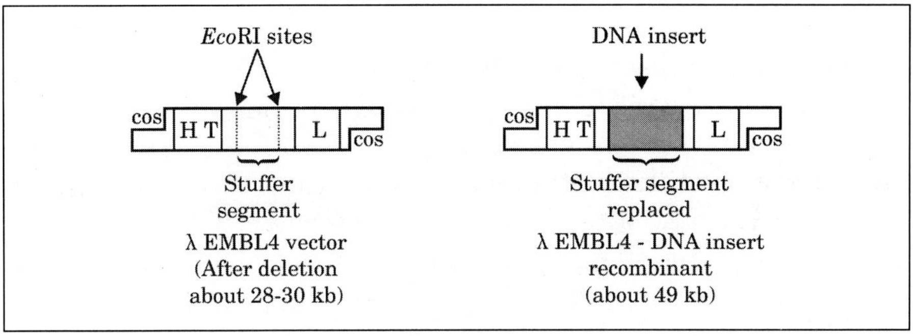

Fig. 5.11: λ Phage based replacement vector λEMBL4 and the recombinant with the stuffer replaced by a DNA insert

Once the phage vector-insertDNA recombinant is formed, it is used for cloning. The insertion into *E. coli* may be done by transfection with the recombinant DNA in the circular form. This method is followed with insertDNA-M13RF recombinants. The process, however, is not very efficient with recombinant λDNA. A better method is to make λDNA-insertDNA recombinant with insertion or replacement vector, in the linear form with *cos* sites and carry out *in vitro* packaging to infect *E. coli*. With this strategy the desired degree of amplification can be achieved.

Cosmids

These vectors are hybrids of parts of bacterial plasmid and λ phage DNA. The structure consists of the bacterial origin of replication, a selection gene as ampicillin resistance gene and a restriction site as *Bam*HI. From the λ DNA only *cos* sequences are taken (Fig. 5.12).

The cos sequences separated by about 40 kb DNA enable *in vitro* packaging of the recombinant DNA. The DNA of interest is inserted in the *Bam*HI restriction gap, and the recombinant subjected to *in vitro* packaging. For cloning, *E. coli* cells are infected with the recombinant phage and plated on agar. After due incubation, because of the bacterial ori in the vector, colonies rather than plaques are formed.

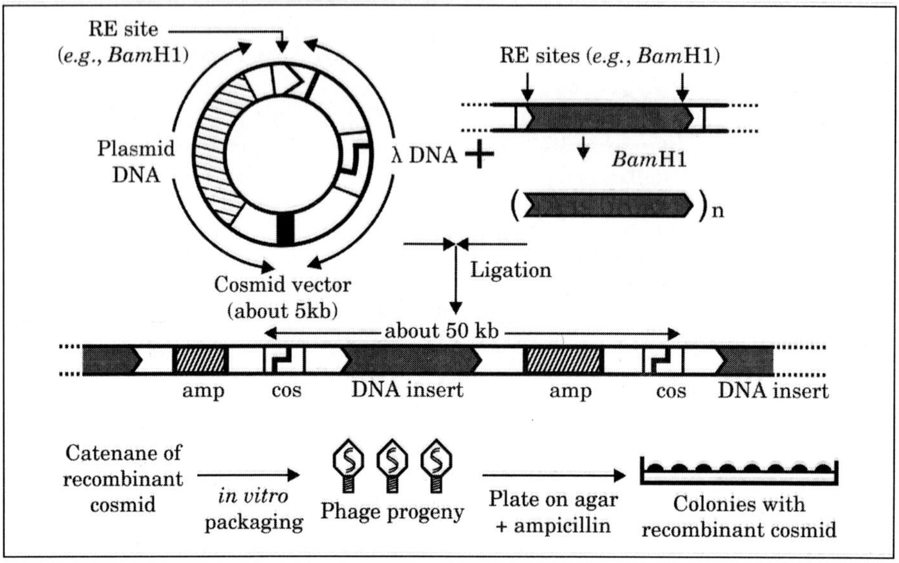

Fig. 5.12: Generalized structure of a Cosmid vector and the recombinant

Some features of the vectors described above are given in Table 5.1.

Table 5.1: Size and carrying capacity of some vectors

Vector	***Size***	***Size of insert (range)***
pBR322	4.36 kb	upto 6 kb
pUC8	2.75 kb	upto 8 kb
m13mp	6.40 kb	upto 3 kb
λgt10	40.00 kb	upto 8 kb
λZAPII	41.00 kb	upto 10 kb
λEMBL4	48.00 kb	upto 20 kb
Cosmids *e.g.* pJB8 (pJB8)	5.40 kb	upto 40 kb
BAC		upto 300 kb
YAC	about 10 kb	upto 1500 kb

Bacterial Artificial Chromosome (BAC)

These vectors are constructed from the F factor (fertility factor) plasmid of *E. coli*. F factors transfer genes across bacteria during conjugation. Various markers and cloning sites are also spliced in

(Fig. 5.13). These can carry DNA inserts upto 300 kb long. Therefore, these vectors have been widely used for genomic studies in both prokaryotes and eukaryotes.

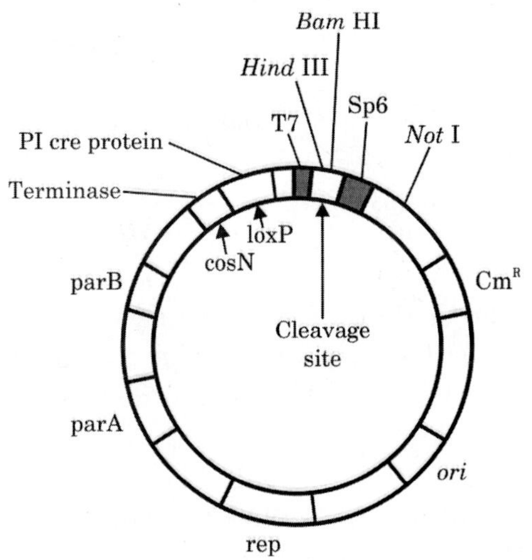

Fig. 5.13: Generalized structure of a Bacterial Artificial Chromosome (BAC) (ori and rep for replication, par A and B for copy number maintenance, Cm^R = chloramphenicol resistance)

5.2. CLONING VECTORS FOR EUKARYOTES

The vectors which are cloned in *E. coli* are useful for many studies in functional genomics. However, for both studies on gene structure, function and biotechnological applications like vaccine production, eukaryotic cells as yeast cells, are used as hosts. For such hosts, some other carrier molecules or vectors are needed. Some useful vectors are the yeast episomal vectors, yeast artificial chromosome (YAC) and the Ti plasmid of *Agrobacterium.*

Yeast Episomal Plasmids

Yeast cells have a plasmid, the 2 μm plasmid. This plasmid exists in about 100 to 200 copies per cell. One of the YEp plasmid vectors,

Yep13 is created by combining the ori region of the 2 μm plasmid, a leucine synthesis LEU2 marker gene from a yeast chromosome and the whole pBR322 sequence (Fig. 5.14).

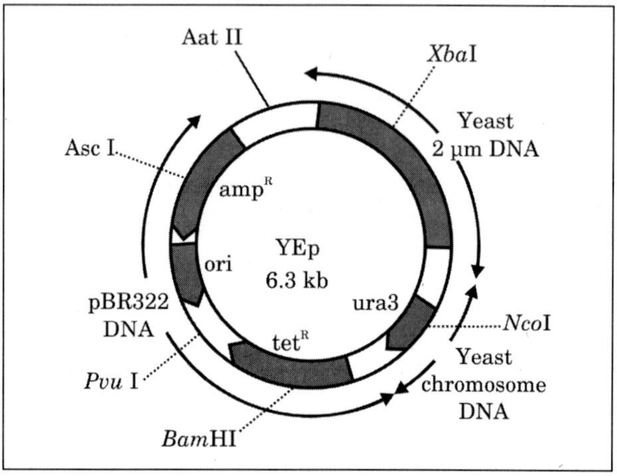

Fig. 5.14: A typical yeast episomal plasmid (Yep) based vector (only few restriction sites shown)

It is a shuttle vector of about 10 kb maximum size, able to replicate both in yeast and *E. coli*. In *E. coli*, selection is by ampicillin/ tetracycline resistance genes. For selection of transformed yeast cells, LEU2 gene acts as the marker. For this, the recombinant DNA is introduced into a mutant strain of yeast, the $LEU2^-$ mutant yeast which cannot synthesize leucine and therefore, cannot grow in agar medium without the amino acid leucine. As the vector has the LEU2 gene, a leucine synthesis gene, the transformed yeast cells become $LEU2^+$ and can survive in leucine minus medium. The DNA of interest is spliced into this vector at one of the restriction sites and the recombinant DNA is introduced into *E. coli* first. Pure DNA insert-YEp recombinant is isolated from transformed *E. coli* culture and then introduced into yeast. A complication is the possible integration of some recombinant molecules into the host yeast chromosome, through the LEU2 containing yeast chromosomal residue of the vector. Attempts are being made to improve the stability of the Yep vector based recombinants.

Yeast Artificial Chromosome (YAC)

The yeast artificial chromosome is constructed by combining segments of yeast chromosome with pBR322. The idea was to create a vector capable of carrying long DNA inserts. The main parts from yeast are the *telomeres* for stability of the vector, the *centromere* for proper partitioning into dividing yeast daughter cells and segments containing *the origin of replication*, besides other segments for selection and insertion of DNA of interest. Thus, for constructing a pYAC, the yeast chromosome segments added to pBR322 are: (i) a TRP1-Ori region, TRP1 being a tryptophan synthesis gene existing in the proximity of the origin of replication sequence, (ii) a CEN4 segment placed after and continuous with the previous segment, being DNA from the centromere region of chromosome 4, (iii) the TEL sequences from the telomeres, with two *Bam*HI sites, (iv) the URA3 gene for pyrimidine biosynthesis and (v) the SUP4 gene, a selection gene to be used for insertion of the DNA of interest, through its *Sna*BI or preferably, introduced *Eco*RI restriction site. The assembly leads to a vector of about 11 kb size (Fig. 5.15).

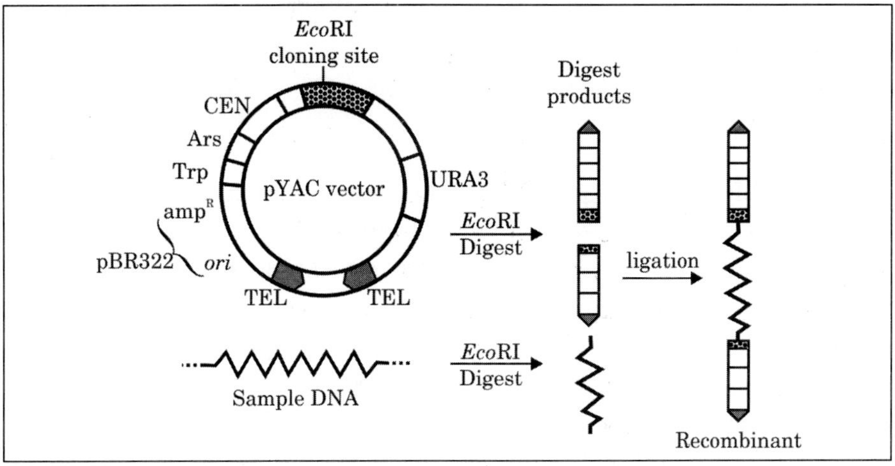

Fig. 5.15: Yeast artificial chromosome based vector and construction of recombinant (amp and ori from pBR322; CEN, Trp, Ars, URA3 and TEL from yeast Chromosome; ori enables replication in *E. coli* and Ars in yeast)

The DNA of interest is cut with *e.g., Eco*R1, mixed with similarly cut YAC vector and ligated into the SUP4 site. This leads to insertional inactivation of SUP4. The DNAinsert-YAC recombinant is introduced into yeast protoplasts. The yeast strain selected is both $trp1^-$ and $ura3^-$, that is incapable of survival in media lacking tryptophan and pyrimidine sources. Transformation with the YAC-insert recombinant makes such yeast cells $trp1^+$-$ura3^+$ and these cells can grow in minimal media while the untransformed ones cannot. The detection of recombinants is done through SUP4 based assay, the recombinants with inactivated SUP4 are white while the others turn red. YAC vectors are very useful for cloning DNA molecules too long for bacterial vectors. Thus, DNA chains of 500 to around 1500 kb can be cloned through these vectors.

5.3. EXPRESSION VECTORS

The vectors mentioned above are used only for DNA amplification through cloning and are therefore, termed as cloning vectors. In expression vectors, a gene is introduced with its promoter and other elements necessary for its expression. Thus, the main use of expression vectors is in studying gene function and to obtain the product of the gene. A brief account of some expression vectors is given here.

Recombinant DNA technology has been extensively used to study the function of genes and other noncoding DNA. This area of study is referred to as functional genomics. For such studies, it is important that the gene inserted in the vector is also expressed. Similarly, for expression of any trait it is necessary that the specifically required protein is produced by the gene insert. For instance, for resistance to boll worm, it is necessary that the Bt gene inserted in a plant cell must produce the Bt toxin protein or a vaccine gene introduced in a bacterium or yeast produces the vaccine. To ensure this, expression vectors or expression cassettes have been designed. The key point is that a gene needs to be expressed in an alien host cell, like a bacterial toxin gene in a cotton plant cell or a fish gene in tomato cells. To make an alien gene express, it is attached in the vector to suitable regulatory elements compatible with the host machinery. The main elements or DNA sequences to be attached are the promoter, ribosome binding and the terminator sequences.

Expression Vectors for *E. coli*

The vectors used are similar to the ones used for cloning except that the above mentioned expression signals are added. Several promoters are used. A common promoter is the *lac* promoter, which normally is the promoter of the *lac*Z gene encoding the β-galactosidase enzyme. The ribosome binding site and the terminator stop signal site are of *E. coli*. The gene of interest is inserted downstream of the promoter and the ribosome binding site which enable expression of the gene. The *lac* promoter is induced by isopropylthiogalactoside (IPTG). Therefore, addition of IPTG to the culture medium of transformed *E. coli,* induces synthesis of the desired recombinant protein.

In certain cases the alien gene is inserted downstream of the promoter within a preexisting gene adjacent to the promoter. This leads to the formation of a fusion protein which is more stable in *E. coli* and is more easily transported out to the growth medium. Many other promoters are used for expression of foreign genes in *E. coli.*

Expression Vectors for Yeast

Yeast (*Sachharomyces cerevisiae*) is a favourite host for the production of recombinant proteins. The gene of the required protein can be inserted in YAC or any other yeast pasmid based vector mentioned above. Upstream of the gene sequence, an yeast promoter is added as also a yeast terminator sequence in the specific $3'$ region of the gene. A commonly used promoter is the *GAL* promoter which switches on the galactose epimerase gene.

Expression Vectors for Plant Cells - *Agrobacterium* Ti Plasmids

In case of plants, the *Agrobacterium* Ti plasmid derived vectors are capable of transferring the transgene, spliced to it, into a plant cell with subsequent integration of the transgene into the host genomic DNA. The plant expression vectors have been described in detail in chapter 8.

Baculovirus Based Vectors for Insect Cells

Animal protein production is difficult in bacteria and yeast as these cells lack the machinery for the final glycosylation of the protein product. Animal cell cultures also do not constitute convenient media because of the need for solid support systems. However, baculovirus based vectors in insect cells are highly efficient for processed animal protein production in good amounts. Baculovirus produces large amounts of polyhedrin protein in the cells of the infected insects. For recombinant animal protein production, polyhedrin gene is replaced by the desired animal protein gene in the baculovirus DNA. Insect cells transformed with such recombinant baculovirus-animal gene DNA, manufacture large amounts of the protein. The commonly used system is of the cells of silk worm (*Bombyx mori*) transformed with a baculovirus, *Bombyx mori*-nucleo-polyhedra virus (BmNPV) containing the required animal protein gene (Fig. 5.16)

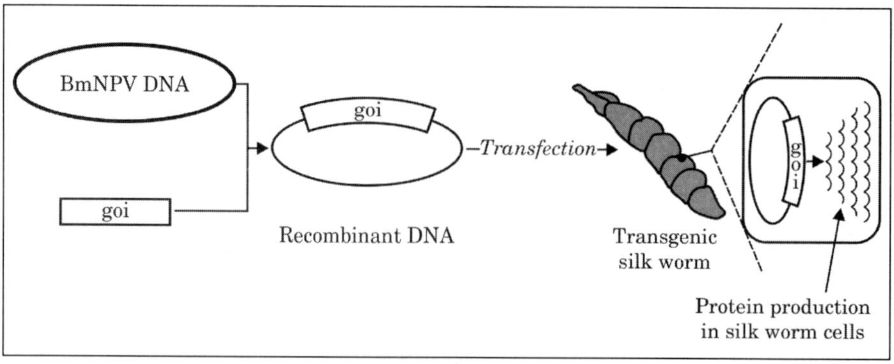

Fig. 5.16: Baculovirus based animal protein production, *Bombyx mori*-nucleo-polyhedra virus (BmNPV)

Some other viruses have also been used for vector designing as, the retroviruses and the adenoviruses. A retrovirus consists of a double stranded RNA genome. On transfection of specially dividing cells, the ds RNA, is converted by reverse transcriptase to ds DNA which integrates into the mammalian genome. Therefore, these can become sources for vector designing. The adenoviruses have ds DNA genome and can transfect nondividing cells. On transfection, the DNA integrates with the host chromosome. Low pathogenecity of adenoviruses to humans make them useful for gene therapy applications.

6

DNA Libraries and Screening

SUMMARY OF CONTENTS

Genomic Library, Complementary DNA (cDNA) Library, Random Arrayed and Ordered Arrayed Libraries; Screening of DNA Libraries-Screening of genes with some available information, Screening of genes on the basis of only phenotype.

A library of books and journals is necessary for students and researchers to learn. Similarly, for genomic studies, it is necessary to have a stock of DNA fragments of genomes of different species, for isolation of genes and studies like gene function, significance of non-coding elements and genome sequence.

6.1. GENOMIC LIBRARY

A collection of individually cloned DNA fragments of the genome of a species is termed as 'genomic library'. A genomic library is prepared by cloning all the DNA fragments of the genome. The number of clones for each genome is vast and depends on the carrying capacity of the vector used as shown below:

Table 6.1: Vector size dependent amplification

Organism	*Genome, bp*	*Colony, no with λEMBL4*[1]	*Colony, no with cosmid*[2]	*Colony, no YAC*[3]
E. coli	1.6×10^6	820	410	–
Saccharomyces	1.8×10^7	3225	1500	–
Human	3.2×10^9	564000	74000	6500

1. Maximum fragment size 17 kb
2. Maximum fragment size 35 kb
3. Maximum size 1400 kb

For obtaining fragments, a genome is subjected to complete or partial digestion by a restriction enzyme. It would be ideal, if each fragment represented the full sequence of a functional unit, like the complete sequence of a gene. Such a condition is difficult to get, with complete digestion, as the genes can get cut and split into different clones. Therefore, partial digestion is also done to obtain larger fragments keeping the genes intact as far as possible. The connection between different DNA fragments obtained from a complete digest cannot be established as these sequences are non overlapping. Partial digests, specially with different restriction enzymes yield, overlapping sequences. From the pattern of overlapping, the sequence of fragments from 5′ to 3′ end can be established. Such studies are of help in *chromosome walking* and also *genome sequencing* as described in chapter 7. The mixture of fragments is mixed with a suitable vector and subjected to ligation. After formation of recombinant DNA in the ligation mixture it is introduced into target host cells for cloning. Plasmid vector-DNAinsert recombinants are introduced into hosts as *E. coli* by transformation or electroporation. Phage vector-DNA insert recombinants are usually packaged *in vitro* and then used for transfecting host bacterial cells. YAC-DNAinsert recombinants are usually introduced into host cells by electroporation. For storage, bacteria from the colonies are pooled and suspended in a medium. Similarly, phages from plaques are collected and pooled (Fig. 6.1).

The tubes are stored at –20°C to –80°C. For cryoprotection, glycerol is added to the bacterial suspension and dimethyl sulphoxide (DMSO) for phages.

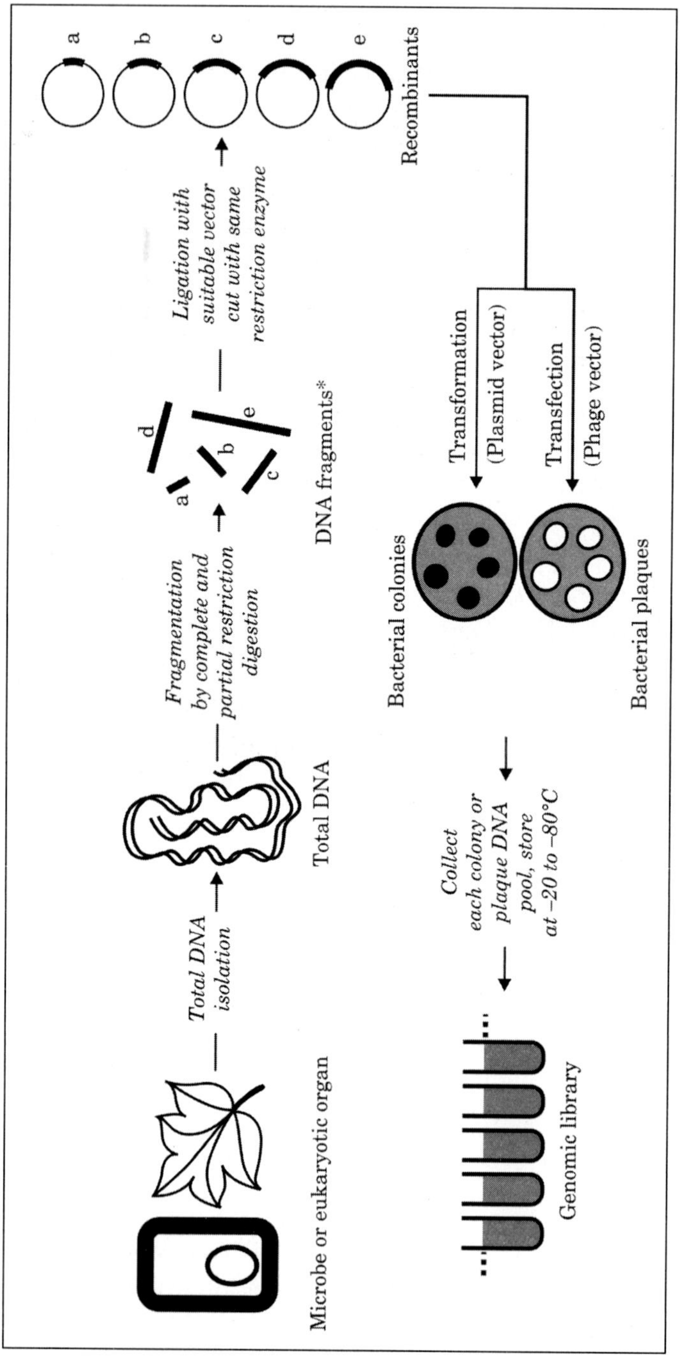

Fig. 6.1: Generalized steps of making a genomic library (*there will be numerous fragments)

To judge the quality of a library, an aliquot of the suspension is digested by the same restricition enzyme as was used for cutting the vector and the DNA products are subjected to gel electrophoresis. Ideally there should be two bands, one of the vector and the other of the insert.

A genomic library is useful specially in cases where the DNA sequence is not known. Further, in case of prokaryotes, which have compact genomes and genes without introns, genomic libraries are convenient. In case of eukaryotes, the interest may be in the exact gene DNA coding sequence without introns. For this, *complementary DNA libraries* (cDNA library) are made with the help of mRNA extracted from an organism. Complementary DNA (cDNA) libraries are also of help in knowing the status of gene expression at a particular developmental stage of a tissue. In multicellular eukaryotes, although the gene complement of all the cells are same, the expression profile of genes vary amongst different tissues and in the same tissue at different developmental stages.

6.2. COMPLEMENTARY DNA (cDNA) LIBRARY

Complementary DNA or cDNA is a *copy* or *complementary* sequence of a strand of messenger RNA. cDNA library is a collection of clones of cDNA synthesized from the total mRNA of an organism or tissue at a particular stage of development. cDNA libraries are of great help in giving an idea of genes active and functioning at a time or developmental stage of a tissue. In contrast to a genomic library, a cDNA library consists of only the genes expressed at a stage and is devoid of other non-coding sequences and introns. To create such a library, first of all, the total mRNA is extracted from a tissue. The mRNA is converted to a cDNA pool by *Reverse Transcriptase*. As mRNAs have polyA tails, poly-dT oligonucleotides being complementary, act as universal primers. Full length cDNAs may also be synthesized by the useful PCR strategy of *Rapid Amplification of cDNA Ends* (*RACE*). The cDNA pool formed is purified by removing RNA with a RNase (as RNase H) treatment. The cDNAs are then

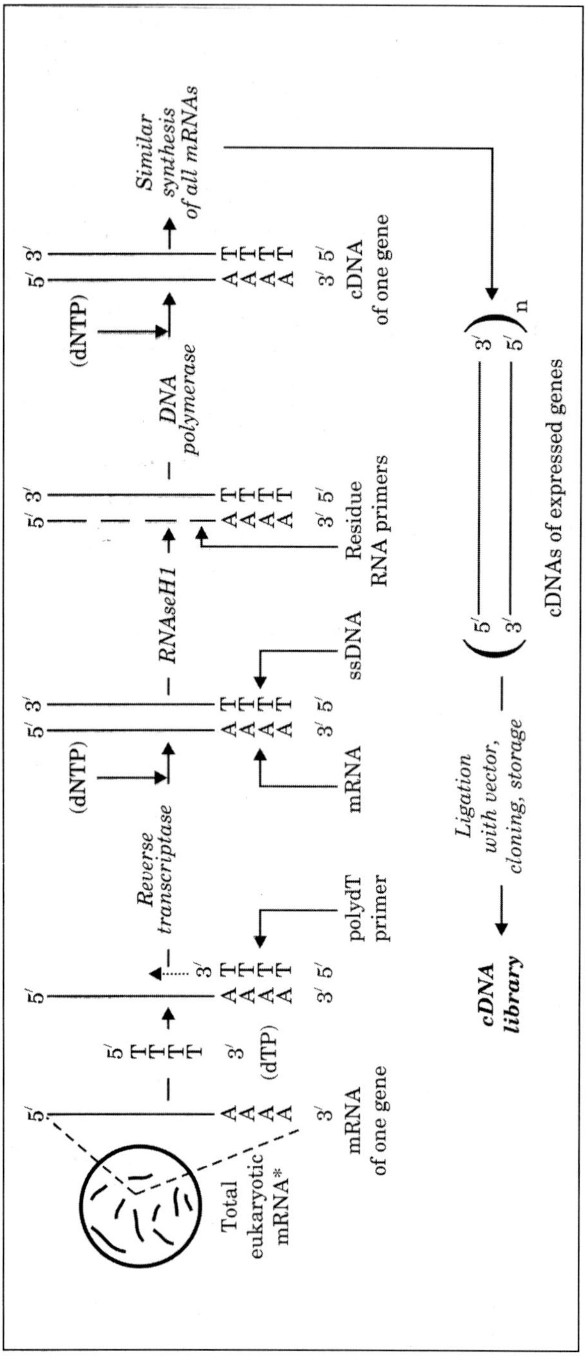

Fig. 6.2: Generalized steps of making c-DNA library(*from organ/tissue/organism)

subjected to usual cloning protocols using mainly phage vectors (Fig. 6.2).

An extension of cDNA libraries is making an inventory of *Expressed Sequence Tags* (ESTs). For this, parts of a cDNA are sequenced by high throughput sequencing techniques. From homology searches partial identification of the protein product and the gene can be made. ESTs are of great help in indicating markers for gene mapping and also give a comparative idea of genes being expressed in different tissues at different times and stages. The latter helps in creating the microarray for transcription studies at global level.

cDNA libraries are mainly useful for eukaryotic systems in which the genes are interrupted by introns and there is a large proportion of nongenic DNA. However, for bacterial systems with compact genomes and absence of introns, genomic libraries can be sufficient. Moreover, long mRNAs of poly cistronic genes and their instability (half life of 1 to 2 minutes) make synthesis of full length cDNAs difficult.

6.3. RANDOM ARRAYED AND ORDERED ARRAYED LIBRARIES

In the above account the genomic library made, exists as a suspension of clones in a tube kept in cold storage. It is also possible through appropriate equipment, to store the bacteria from different colonies or phages from different plaques separately in a row or array of wells. There are two types of arrays. In *random array* the sequence of the wells is not in the order of the genomic sequence. But, by finding consecutive overlaps in fragments, it is possible to store the recombinant bacteria/phages in an *ordered array* similar to the linear sequence of the genomic DNA.

6.4. SCREENING OF DNA LIBRARIES

Genomic libraries contain clones of DNA fragments constituting genes, regulatory sequences as also nongenic regions. The prime interest is in identifying genes controlling different phenotypes out of the numerous available clones. Such detection of the clone of a gene in a library is called *screening*. The methods are based on two situations, a) there is some idea of the sequence on the basis of the protein product and phenotype and b) the phenotype is known but there is no idea of the sequence. Strategies for both situations are briefly described below.

Screening of Genes with Some Available Sequence Information

Some of the approaches are described in the following sections.

Direct screening

There are several detection methods based on *expression*. The traits chosen can be like antibiotic resistance, dependance on a metabolite for growth or production of an enzyme, the product of which gives a colour test. The library suspension is plated and specific colony/ colonies displaying the trait is selected. For tests based on a *visible phenotype,* a classic example is the gene coding for catechol 2,3 oxygenase in *Pseudomonas aeruginosa*. This enzyme converts catechol to a yellow product. Thus, in library plate, the colonies turning yellow are the ones with the gene. Direct selection of a gene, for instance an antibiotic resistance gene in *E. coli* is shown in Fig 6.3.

Complementation or marker rescue

The marker rescue approach has been found to be quite useful. Complementation is seen as the recovery of a phenotype in a mix of two genes, where each gene has a part control for expression of a phenotype. This principle has mainly been used for screening of genes involved in the biosynthesis of certain metabolites, specially organic nutrients. For example, to search for a gene for lysine biosynthesis, the *E. coli* lys^+ library suspension can be introduced in an auxotrophic *E. coli* lys^- mutant unable to synthesize lysine.

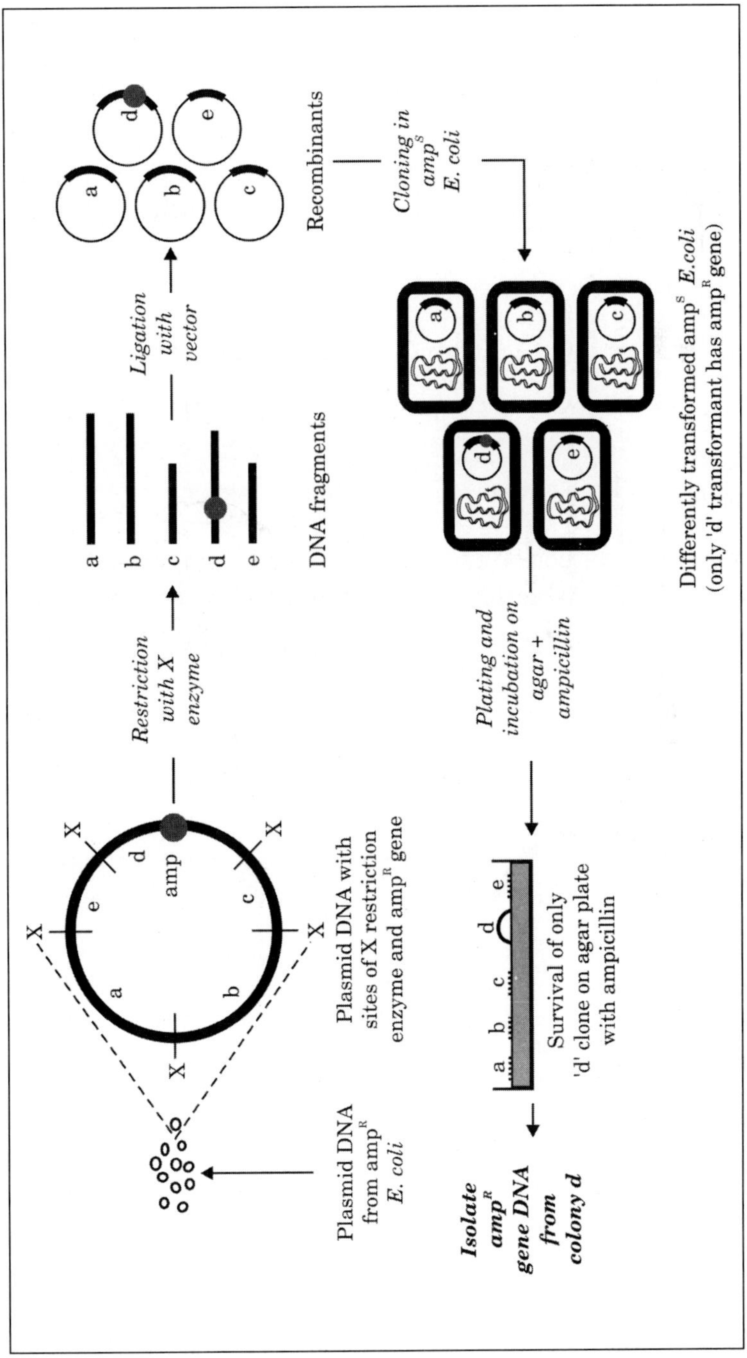

Fig. 6.3: Direct screening of amp^{R} gene in *E. coli* (hypothetical example)

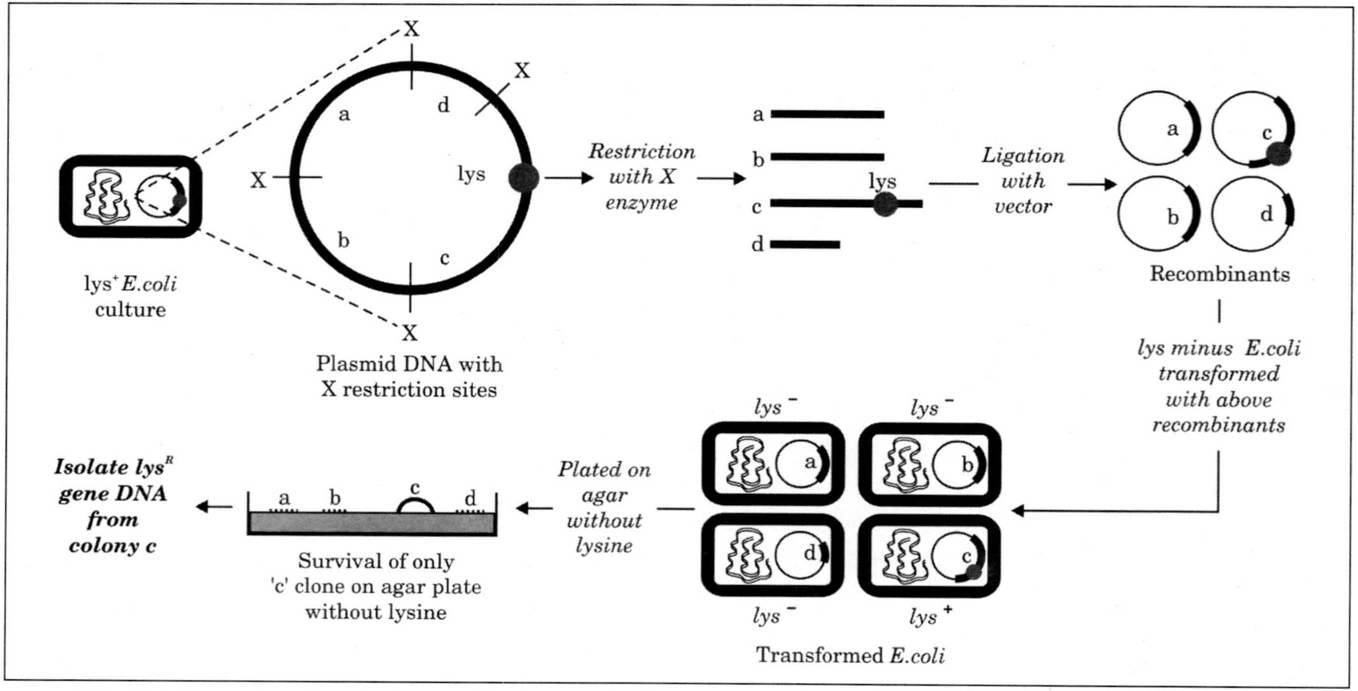

Fig. 6.4: Screening for a lysine synthesis gene, lys$^+$ gene by marker rescue method (hypothetical example)

On plating of the transformation mixture on a lysine minus medium, colonies of bacteria which can grow, contain the gene (Fig. 6.4).

Immunological methods

These methods can also be used. In the case of λ phage vectors, the entry of the recombinant DNA into bacteria causes lysis, resulting in release of cellular proteins in the plaque. Each protein can be traced back to a gene. For immunoassay, the antibody against the known protein product of a gene is raised. The proteins from the plate of plaques is lifted and transferred to nitrocellulose filters, in the same pattern as the original plate (as in replica plating). The proteins get bound to the filter and are washed with the antibody solution. The antibody recognizes a short sequence (linear *epitope*) or specific conformation (conformational *epitope*) of the protein. Specific protein-antibody binding takes place and is detected. The plaque/s showing such binding contain/s the gene.

Hybridization of sample DNA with complementary labeled probes

For *DNA sequence* based screening, labeled oligonucleotide probes are used. Short synthetic oligonucleotides of about 20 bases, like the primers of PCR are made. The designing of a probe for a gene of interest is based on the known sequence of its protein product or sequences of homologous proteins from other species. A short sequence for instance, from the N terminal end of a protein is chosen. All possible labeled oligonucleotides are synthesized or *back translated* taking into consideration all alternate codons. Probing can be done by various techniques as Southern blot, dot blot or zoo blot. For probing of genomic library components, cDNAs from the cDNA library of the same genome can also be used. Hybridization can be performed in solution, on filters or *in situ*. The probes are mainly labeled with P^{32}, P^{33}, S^{35} and H^3. Radio isotopes as P^{32} are better for protocols as Southern blotting and H^3 for detection through *in situ* hybridization. In recent times, non-radioactive probes are

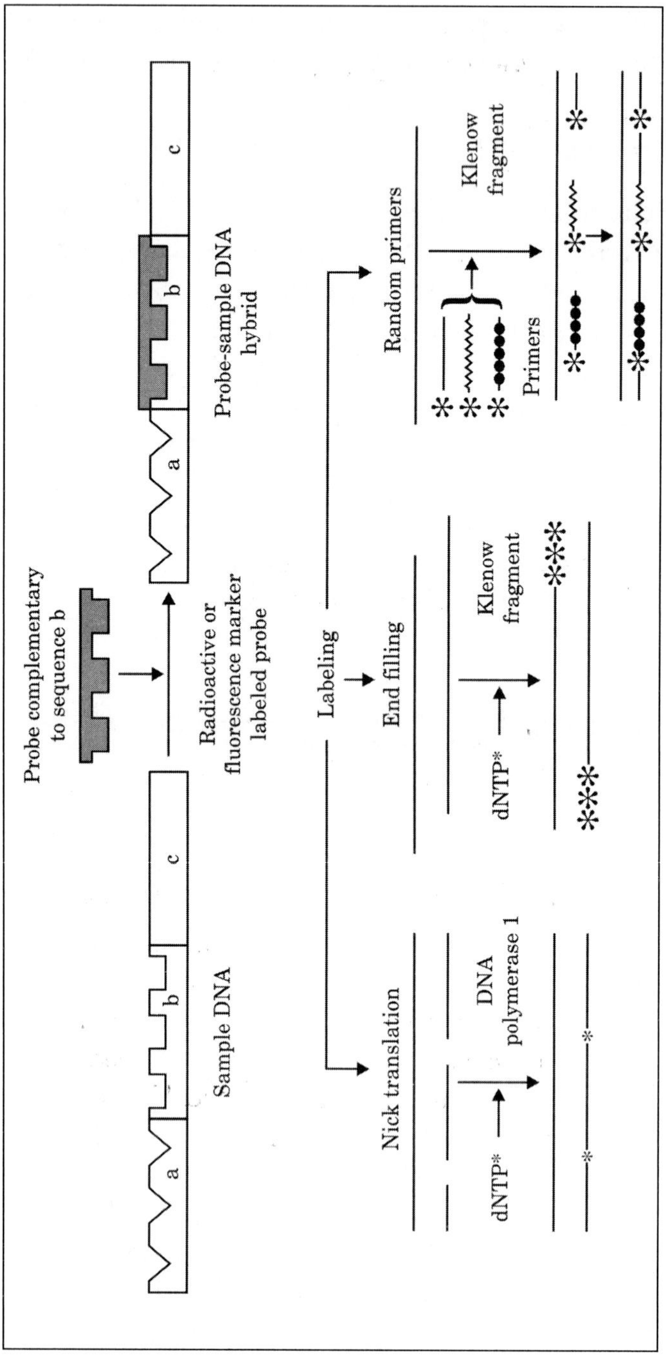

Fig. 6.5: Principle of probing and some methods of labeling probes (*radioactive isotope *e.g.*, P^{32} or fluorescence marker)

becoming more popular as, they can be stored for longer periods without decay unlike the commonly used radio-isotopes with short half lives. Some main ways of labeling probes are shown in Fig. 6.5 and detection of hybrids in Fig. 6.6.

Probes are mainly labeled by the protocols of nick translation, end filling or random priming.

The radio labeled hybrids are detected by autoradiography and fluorescence markers by fluorimetry.

Screening of Genes on the Basis of Phenotype Alone

Searching for some gene without any biochemical information and only on the basis of just phenotype is quite a challenge. However, some innovations have helped.

Gene tagging with transposons

Transposons are mobile elements occurring in most genomes. On getting inserted in or adjacent to a gene, they mutate the gene leading to a loss of its phenotypic expression. If the transposon contains a marker, it is even better. After transformation and the required transposition of a cell, the mutant can be recognized by the loss of the phenotypic trait. More importantly, the gene mutated because of linkage with the transposon and as the transposon has a marker, the gene can be located, that is the gene has got *tagged*. Transposons are basically, DNA segments in a genome which can be removed by a transposase enzyme from their original site to another location in the genome. The transposase gene may occur within the transposon sequence (*autonomous* transposon) or in another region of the genome (*nonautonomous* transposon). Further, a transposon may be replicated (*replicative* type) and the copy inserted elsewhere or is removed as such (*nonreplicative*) and inserted at some other location. Selection of a transposon for mutating and tagging a gene, can vary for different situations.

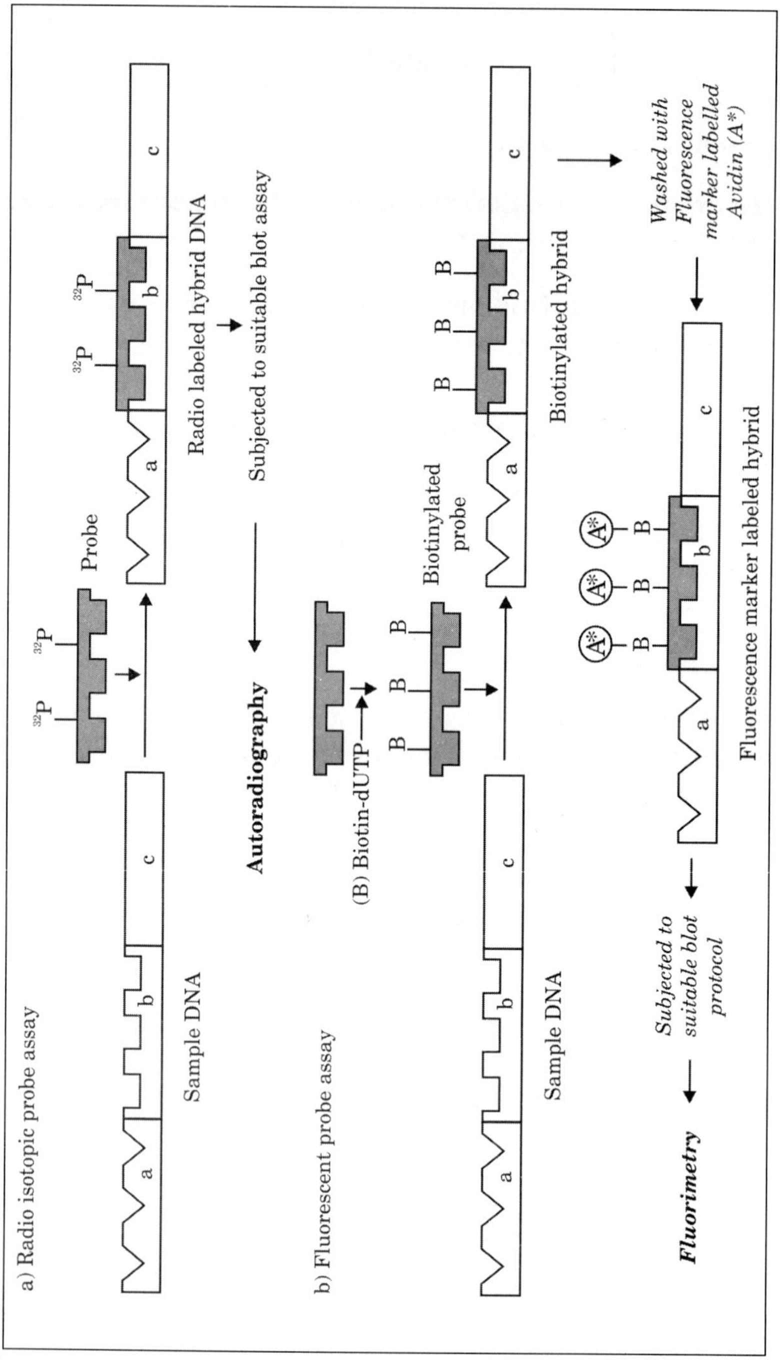

Fig. 6.6: Detection of radio isotopic and fluorescent hybridized probes

One basic approach is to use preferably, an autonomous transposon with a marker gene, two origins of replication, one for *E. coli* and the other for the target cell. The transposon is ligated to a plasmid or viral vector (without ori) and introduced into a host cell containing the gene of interest. On entering the cell the transposon jumps from the vector to the host genome. In practice the vector-transposon recombinant is incubated with the host cells. Therefore, there would be many transformed cells containing the transposon in their genome. In some cells the transposon will attach itself to the gene of interest, causing its mutation and in some it will attach to other genes. Cells which show loss of the phenotypic trait are selected. Total DNA is extracted from the cells, subjected to restriction digestion ensuring that the transposon remains untouched and a genomic library is made. The colonies formed are replica plated on a medium containing the antibiotic specific for the antibiotic resistance marker in the transposon. The colonies resistant to the antibiotic only survive. The marker is present on the transposon and the transposon is linked to the gene of interest. Therefore, this clone represents that functional gene. The DNA is extracted from this clone and sequenced. As the transposon sequence is known, the rest of the sequence of the DNA extracted is of the gene.

Signature tagged mutagenesis

This is an ingineous method for identifying genes involved in complex processes. It is again based on transposon mediated mutagenesis. The transposons are however special, having being assigned identification or *signature* in the form of attached unique sequences. The method was developed in investigations for identifying genes involved in the fatal infection of mice by *Salmonella typhimurium*. The method is briefly outlined here. A set of oligonucleotides, 80 bases long, are synthesized with same 20 base $5'$ and $3'$ ends. The mid 40 base region sequence is varied, thus constituting a *signature tag* for each transposon. Each oligonucleotide is inserted at a restriction site of the transposon *miniTn5km2*. This transposon resides on a plasmid capable of replicating in *E. coli* but not in

Salmonella. The transposon *miniTn5Km2* is a modified Tn5 transposon with a kanamycin resistance gene added and the transposase gene shifted to the host plasmid. Removal of the transposase into the plasmid which cannot grow in *Salmonella* ensures that only one transposition event will take place. The plasmid-transposon recombinant is introduced into *Salmonella.* After the vector-transposon recombinant enters a *Salmonella* cell, the transposon jumps to the bacterial genome and the plasmid with the transposase, incapable of replicating, is unable to pass on to the daughter cells. The transformation mix is plated on kanamycin containing agar. On this plate, only the colonies of bacteria transformed by the signature tagged transposon with kanamycin resistance, can grow. All the colonies are now transferred to an array of well plates with one colony in each well. The array is replica plated on an agar plate and once the colonies grow, again these are replica tranferred to a nylon filter. The filter is placed on agar for the bacteria to grow and then from each, DNA is released. Thus, one has a filter plate of an array of DNAs from different transformed colonies, a mirror image of the original plate. *Each of these colonies has DNA with the transposon inserted in different positions and only some with the transposon inserted in an infection inducing gene.* While colonies with transposon inserted in other genes can infect mice, the colonies with transposon in the 'infection' gene, that is with the mutated infection gene cannot. To find out such colonies, mice are infected with bacteria from each well. After few days, bacteria are taken from the spleen of the mice and grown. Obviously, the bacteria with intact infection genes infect mice organs and survive and grow there. The ones with the mutated gene cannot infect and so, cannot survive. DNA is extracted from each surviving colony, sequence tags amplified and used as probes for hybridization with the original nylon filter plate. The gaps indicate positons of infection genes. By comparison with such gap positions, the mutated infection gene DNA, attached to the sequence tagged transposon is identified, extracted and processed for other studies.

Transposon tagging for gene identification has worked successfully in the case of plant genes also. A classic example is the disease resistance gene cf9 of tomato which counters the infection of the fungus *Cladosporium fulvum.* On infection by *Cladosporium* the cf9 gene is expressed in the infected tissue resulting in the release of a protein. This protein combines with a protein produced by the avirulence, *avr* gene of the fungus. This interaction leads to death and drying of the infected tissue leading to a check in the spread of the fungus. For tracking the cf9 gene, two tomato lines were developed, one homozygous for the *avr* gene and the other homozygous for the cf9 gene. In the cf9 homozygous line, the transposon Ds containing an antibiotic resistance gene and *E. coli* ori was introduced in the chromosome containing cf9 and the transposase Ac element on another chromosome. On crossing, the main progeny lines contained both *avr* and cf9 genes and some with avr and Ds associated mutated cf9. The former progeny died because of *avr* protein and cf9 protein interactions. The latter survived as only avr protein was formed and the Ds-cf9 was non functional. From these plants the DNA was isolated, subjected to cloning, the clone containing Ds-cf9 screened out by antibiotic selection. From this the Ds-cf9 DNA was isolated and studied.

7

Genome Components and Sequencing

SUMMARY OF CONTENTS

Genome components; Genome sequencing strategies; Updating chromosome maps; Assembling sequence reads into contigs; Sequencing of the whole genome; Some plant genomes sequenced; Some milestones in the history of Human Genome Project.

This chapter starts with an overview of the components and organization of the genome. Such a general description is followed by an account of strategies of genome sequencing which was catalyzed by the Human Genome Project.

7.1. GENOME COMPONENTS

The chemical basis of heredity in all organisms, prokaryotes or eukaryotes, is nucleic acid. In the majority of species it is DNA and rarely in some cases, as the retroviruses, it is RNA. The total nucleic acid of an organism constitutes the *genome*. The DNA of a cell is in the chromosomes. In eukaryotes the chromosomes are delimited within the nucleus but in prokaryotes they are not delimited by a membrane envelope and remain exposed. Further, in all eukaryotic cells DNA is also present in the mitochondria and additionally in plants, also in chloroplasts. The organellar DNA is very small in size as compared to the nuclear DNA. In the case of multicellular eukaryotes, there are the somatic cells which make up the body and the male and female sex or germ cells in the sex organs. The somatic cells are diploid and the germ cells are haploid. Therefore, the quantity of DNA per cell or the C-value is expressed as the amount of DNA for the haploid set of chromosomes. The length of DNA per cell, ranges from approximately 3000 base pairs (3×10^3 bp) of some viruses to 100000000000 bp (10^{11}bp) in some flowering plants. Studies of genome organization revealed that the DNA content of most organisms, specially the eukaryotes, is much more than the amount estimated on the basis of genetic activity or the probable number of genes. This was referred to as the *C-value paradox*. Such observations led to detailed studies on undertstanding the sequence complexity of the genomes of different organisms. Two types of initial studies done in late sixties, *reassociation kinetics* which involves studies on reassociation of the single strands of denatured synthetic and natural DNA, and *equilibrium density gradient centrifugation* revealed important aspects of the sequence complexity of genomes.

The rate of reassociation of single stranded complementary DNA, separated by heat and then allowed to anneal, is related to the extent of complementarity between the two strands. The strands anneal faster if the proportion of repeating units is higher. It was shown in 1960s that when DNA is heated to its Tm (temperature for 50% melting) and allowed to reassociate at 25°C lower temperature, the rate of reassociation can be estimated by the relationship:

$C/C_0 = 1/1 + k.C_0t$, where C_0 is the concentration at the start (time t = 0), C is concentration at a particular time t and k = reassociation rate constant.

At 50% reassociation the above relationship becomes $C_0t_{1/2} = 1/k \text{ mol} \times s^{-1}$

Thus, the rate of reassociation will decrease with increase in the initial concentration. Moreover, the reassociation rate depends on the extent of complementarity between the two strands. Thus, homopolymeric strands like ..AAAAAAAA... and ...TTTTTTTTT.., will reassociate faster and have lower C_0t value than ...ATGCGCAT.. and ..TACGCGTA... . An extension of this logic is that sections of a genome with repeating units will reassociate faster and will have lower C_0t values than sections with unique sequences. This situation was found to be true. Various experiments with prokaryotic and eukaryotic whole genome DNA fragments indicated prokaryotic genomes to consist of mainly unique sequences and the eukaryotic genomes to have mixes of repetitive and unique sequences (Fig. 7.1).

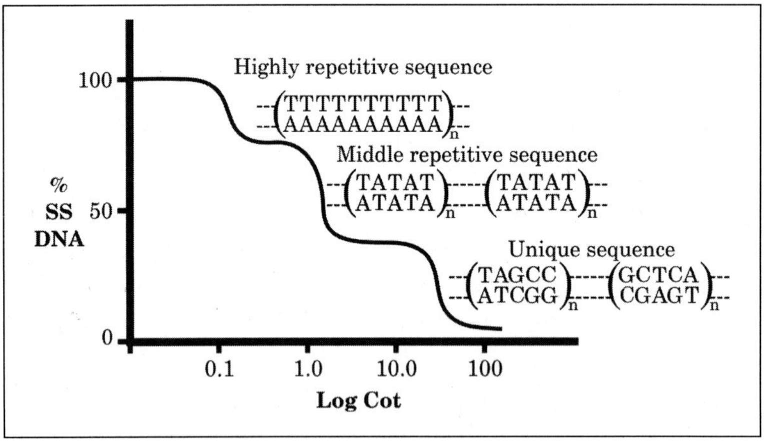

Fig. 7.1: Reassociation of different types of single stranded DNA

Density gradient equilibrium ultracentrifugation also revealed the occurrence of repeat sequences in eukaryotic DNA. When total DNA was subjected to caesium chloride density gradient ultracentrifugation, besides the main band, some minor bands were also obtained. The main band equilibrates to its own buoyant density of around 1.7 [p or density = 1.66 + 0.00098 (% GC) g cm^{-1}] and some minor *satellite* bands at slightly lower or higher buoyant densities (Fig. 7.2).

Sequencing of the satellite DNAs obtained, showed them to mainly have repeating sequences and specially tandem repeats.

Various studies indicated the ubiquitous occurrence of the following components in eukaryotic genomes.

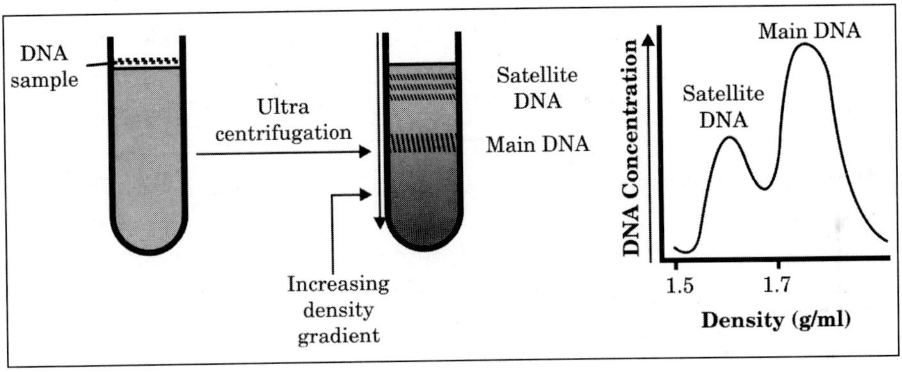

Fig. 7.2: $CsCl_2$ equilibrium density gradient profile of eukaryotic DNA

Components of the eukaryotic genome

1. Genes
2. Gene related elements
3. Repeat sequences:
   a) Tandem Repeat Sequences:
      Minisatellites (VNTRs) and Microsatellites (STRs)
   b) Dispersed or genomewide repeat sequences
      LINES (long interspersed nuclear elements)
      SINES (short interspersed nuclear elements)

   Some LINES and the SINES are transposable elements or retrotransposons

   LTR elements (long terminal repeat elements)
4. DNA transposons

   Genes in eukaryotes are interrupted and in prokaryotes uninterrupted (Fig. 7.3). Every gene has a row of coding sequences the *exons* interrupted by noncoding regions the *introns.* Usually the length of introns is much more than the exons.

The related elements for a gene consist of the *trans* enhancers, UTRs (un-translated regions) and some other sequences as the *pseudogenes*. Genes have been the traditional *markers* for

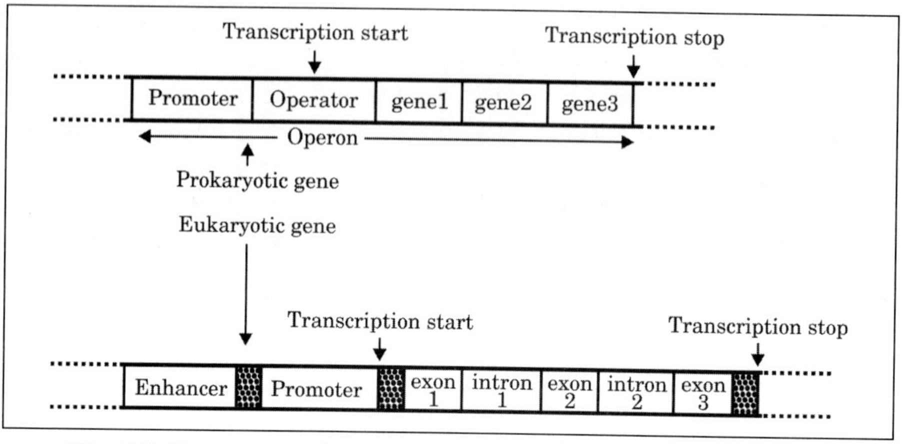

Fig. 7.3: Uninterrupted prokaryotic and interrupted eukaryotic gene

chromosome mapping. Genes became useful as markers, as, each gene occurs in multiple forms, the alleles, that is, are *polymorphic.*

The *minisatellites* or VNTRs (variable number tandem repeats) are sequences containing consecutive repeats of about 10 to 25 bp sequences. *Microsatellites*, also referred to as STRs (short tandem repeats), consist of tandem repeats of 2 to 10 bp sequence elements. Both mini and micro satellites are highly polymorphic and have been extensively used for genetic and physical mapping of chromosomes.

The dispersed or genome-wide repeat elements consist of LTR elements, LINES and SINES.

The LTR elements are retrotransposons with repeats in the two end sequences. DNA transposons are mobile elements which move within a chromosome from one region to another or between chromosomes. The movement can be by excision of the element from one region and reintegration into another (conservative mobility). Another way is the replicative mode. In this mode, the transposon DNA replicates and the DNA copy then integrates into a new chromosomal site. These two methods occur with DNA transposons (item 4 above). LTR elements, SINES and LINE type-1 are retrotransposons. In these cases, the transposon template transcribes to form a complementary RNA which is reverse transcribed to DNA.

This DNA copy then reintroduces to another region. As this pathway of integration resembles the mode of RNA retroviruses, these transposons are termed as retrotransposons. Retrotransposons are much more common in eukaryotes than in the prokaryotes. In contrast, the DNA transposons are more common in prokaryotes. The first transposons, the AC/Ds mobile 'controlling elements' of maize were discovered by Barbara McClintock in 1950s. Her discovery and concept were accepted only in 1970s. The proportion of various componenets in human genome are shown in Table 7.1.

Table 7.1: Proportion of the various components in the human genome

Component	*Size, Mb*	*Percent/Total*
Genes		
Exons	48	1.5
Related	1152	36.0
Microsatellites	90	2.85
Minisatellites		
etc.	510	16.0
LINES	640	20.0
SINES	420	13.0
LTR	250	7.8
DNA transposons	90	2.85
Total size	3200	–

7.2. GENOME SEQUENCING OVERVIEW

As described above, genome is the total DNA of a cell in a species. Each cell consists of DNA molecules, one in each chromosome. Genome sequencing of a species involves determination of the base sequence from $5'$ to $3'$ end of the DNA of each chromosome. Thus, genome sequence of a species is the sequence of the DNAs of all its chromosomes. The sequence shown is of the strand which has the sequence of codons and not of the complementary template strand.

The approaches to sequencing may vary with the complexity of the genome. For instance, the protocol used for a prokaryotic genome with uninterrupted genes and few repeat sequences, may not work in the case of an eukaryotic genome, with genes having coding exons and noncoding introns and also many repeated sequences. Although sequencing efforts started with successes in sequencing of a t-RNA and a small virus, the Human Genome Project induced a quantum jump in sequencing science. It catalysed rapid advances in relevant technologies and also laid the principles of joint effort and open access information in the pursuit of science. Therefore, in the following sections there is greater emphasis on human genome sequencing.

The human genome consists of 24 chromosomes, 22 pairs of autosomes, the two sex chromosomes X and Y and the mitochondrial chromosome. For sequencing purposes, the haploid set is of concern. This is because the linear DNA molecules are similar in both members of an autosome pair and only in the X, Y and the mitochondrial chromosome are different. The 25 DNA molecules of the haploid set add upto about 3.1 billion (3.1×10^9) nucleotides. The sequencing could be done because of advances in technology as: (1) possibilities of DNA amplification by recombinant DNA based gene cloning and PCR (polymerase chain reaction), (2) availability of, besides genes, many polymorphic sequence markers and sequence features, (3) high through-put automated sequencing methods for DNA fragments of about 800 bases length, (4) bioinformatics involving softwares for organizing sequence data and algorithms for assembling sequence data and gene annotation.

7.3. GENOME SEQUENCING STRATEGIES

The basic generalized strategy of sequencing the genome of an organism entails the following studies:

1. Updating genetic and physical maps of the chromosomes of the species as reference.
2. Extraction of chromosomal DNA.
3. Fragmentation of the DNA into large fragments and cloning in high capacity vectors as BAC.
4. Sequencing each clone insert to obtain *sequence reads*.
5. Conducting studies on clone fingerprinting, including RFL patterns, and STS content.

6. Uploading the data into a computer algorithm programme which processes and assembles the overlapping sequence reads in a linear contiguous arrangement or *contig* and the contigs also into a linear order, on the basis of overlaps and available genetic and physical map data.
7. To fill the inevitable remaining gaps, each large insert is further broken into smaller fragments and subcloned in a suitable vector. The sequences of either end of each insert is determined to obtain *paired end reads*. Some other studies are also conducted and a computer algorithm again assembles and refines the sequence further.
8. The above studies lead to the construction of a *draft* sequence. Arriving at the *finished* sequence, may need repetitions of above studies and additional experiments. However, advances in technology as automated sequencing, robotics and improved algorithm softwares have made the task easier, faster and cheaper. The basic steps are outlined in Fig. 7.4.

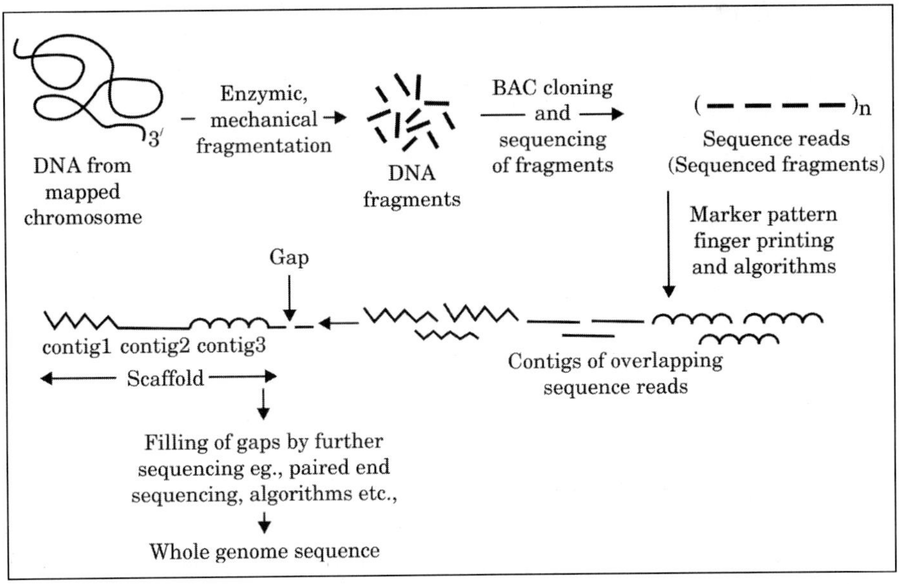

Fig. 7.4: Generalized steps of genome sequencing

It must be noted that the accuracy of any sequencing programme depends on how extensively its sections have been sequenced and the number of times the bases have been represented in the sequences. This is indicated by the number of times a nucleotide of a genome appears in the sequence reads, that is, the *coverage* of the technique. For accuracy this is estimated by the formula,

Coverage = N × L/G, N = number of reads, L = average read length and G = genome size.

Thus, for a 10000 bp genome (G) with 100 reads (N) and average length 1000 bp (L), coverage is 10x (for accuracy and more than 90% representation, the coverage should be ten times of the sample DNA length).

7.4. UPDATING CHROMOSOME MAPS

This involves genetic mapping and physical mapping.

Genetic Mapping

Genetic maps are constructed by genetic techniques. They have been classically made with the help of genes of distinct phenotypic traits and in recent times with the help of some unique DNA sequence sets, which help as markers. Markers that are useful in genetic or physical mapping are those which occur in different forms as alleles, that is, are multiallelic or polymorphic. Further, the markers must be well distributed along the genome.

Genes as markers

In all classical studies the alleles of genes served as markers. Even now genes serve as markers. In this technique series of crosses are made between parents with contrasting alleles of a trait and inheritance patterns analysed. Genetic maps are constructed mainly by observing the recombination of genes during meiosis for gamete formation and the inheritance pattern of the genes. The genes occur in the DNA of each chromosome. In the experiments of Mendel, the advantage was that the genes of different traits occurred on different chromosomes and the alleles of each trait occurred on different chromosomes of a homologous pair. This led in a monohybrid cross, to a segregation ratio of 3:1 of dominant: recessive traits in the F_2 progeny. In dihybrid crosses the ratio was 9:3:3:1. So, the genes were independent and unlinked. If the genes being studied are unlinked and have clear dominant and recessive nature, their segregation in future progeny would be in this classic Mendelian pattern. Several crossing experiments had indicated that many a time this independent segregation and assortment of genes do not happen.Thus in some dihybrid crosses a ratio of 3:0.3:0.3:1 was obtained. Such situations were explained by considering and demonstrating that genes can be linked (Fig. 7.5) and therefore, do not segregate and assort independently.

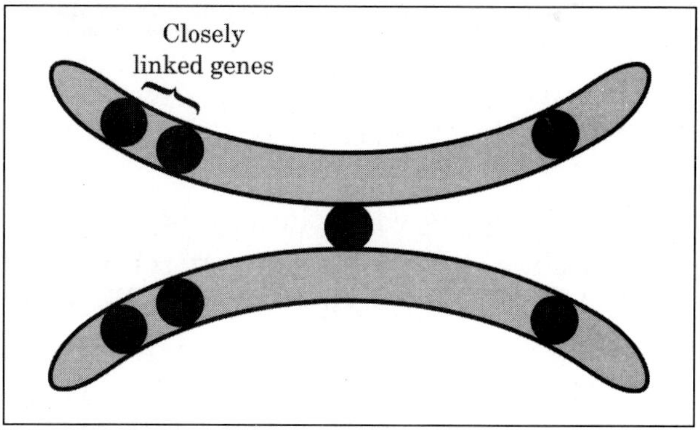

Fig. 7.5: Linked genes

Morgan and his students had demonstrated that genes occur on chromosomes. For gamete formation, the progenitor sex cells undergo meiosis which allows each gamete to receive a haploid set of chromosomes. The homologues of a pair of parental chromosomes may reach the gametes as such or after some modification. The modification involves exchange of parts between the homologues of a pair of chromosomes during the crossover phase of meiosis. Crossing over leads to formation of slightly different chromosomes of a homologous pair due to recombination of some genes. Recombination means some genes coming together and some separating from each other, within a pair of homologous chromosomes (Fig. 7.6).

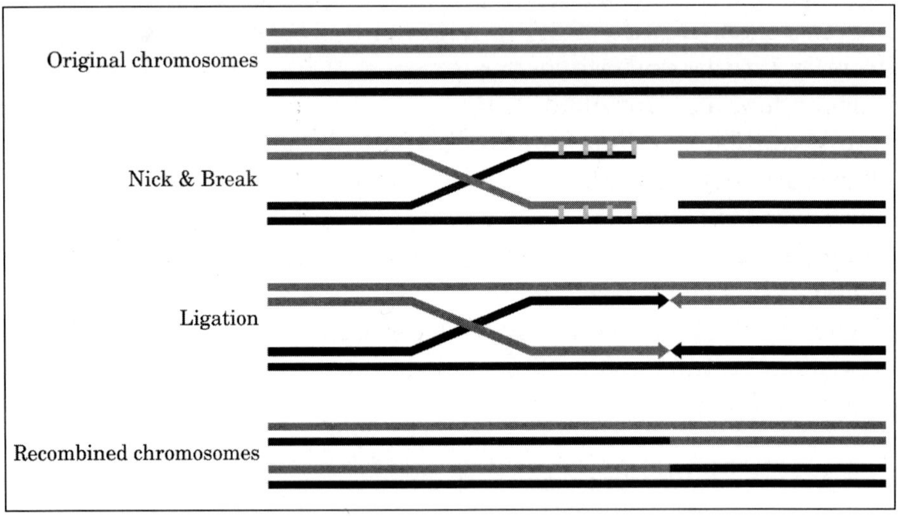

Fig. 7.6: Recombination of genes by crossing over between homologous chromosomes during meiosis.

Various designed crosses with plants and animals have shown that, genes which show high frequency of recombination during crossover in meiosis are far apart in a chromosome. Similarly, genes with low frequency are near each other. The unit of distance between genes is cM (centi Morgan, after the great geneticist Thomas Morgan). A recombination frequency of 1% is 1 cM and has been found to be equivalent to a length of approximately 1.2 million bp (1.2 Mb) through physical mapping. In plants and animals it is possible to determine linkage and recombination frequencies by designed crossing. Conducting test crosses, with selected contrasting traits and through estimation of recombination frequencies, genetic maps of various chromosomes have been constructed. There are, however, some limitations of genes as markers. In eukaryotes the genes are a small component of the genome, are far apart and many are not multiallelic. Therefore, the discovery of special DNA sequences as repeat units and other heritable variations in the sequence have been very useful.

DNA sequences as markers

Recombinant DNA technology ushered in a new era in DNA science. Various sections of the genic and nongenic DNA could be isolated, amplified, manipulated and studied. Many components of genomic DNA had been identified by mid 1980s as mentioned above. For sequencing, of special interest were the sequences which existed in different forms, as alleles, that is, exhibited sufficient polymorphism. These sequences could be used as *markers* for arranging the different components of the genome in a correct order from one end to the other, in the $5' \rightarrow 3'$ direction.

The commonly used DNA sequence components which serve as markers are enumerated below.

1. Restriction site variations leading to *restriction fragment length polymorphisms* (RFLPs).
2. Simple sequence length variations resulting in *simple sequence length polymorphisms* (SSLPs) comprising the minisatellites (VNTRs) and microsatellites (STRs),
3. Single nucleotide changes at particular sites due to point mutations with subsequent heritable single nucleotide polymorphisms (SNPs).

All these are variants of a sequence, multiple forms of a sequence, like the alleles of a gene.

Genes are detected usually by the visible phenotypic traits they control. These traits whether morphological or biochemical can be easily traced. The sequence markers on the other hand do not control any traits and hence have to be detected by some biochemical methods. They need special techniques for detection.

RFLPs

Restriction fragment length polymorphism is due to alteration in the recognition site of some restriction enzyme. In one chromosome/ organism an intact recognition site may exist and in another, the same site may be modified by a single nucleotide change. In the former case, a DNA fragment is degraded by a restriction enzyme resulting in two fragments and in the latter case the enzyme does not work, thus leaving the fragment intact (Fig. 7.7). RFLPs can be detected by gel electrophoresis followed by Southern blotting. However, a better method is the use of PCR with primers corresponding to known sequences flanking the restriction site, followed by gel electrophoretic separation.

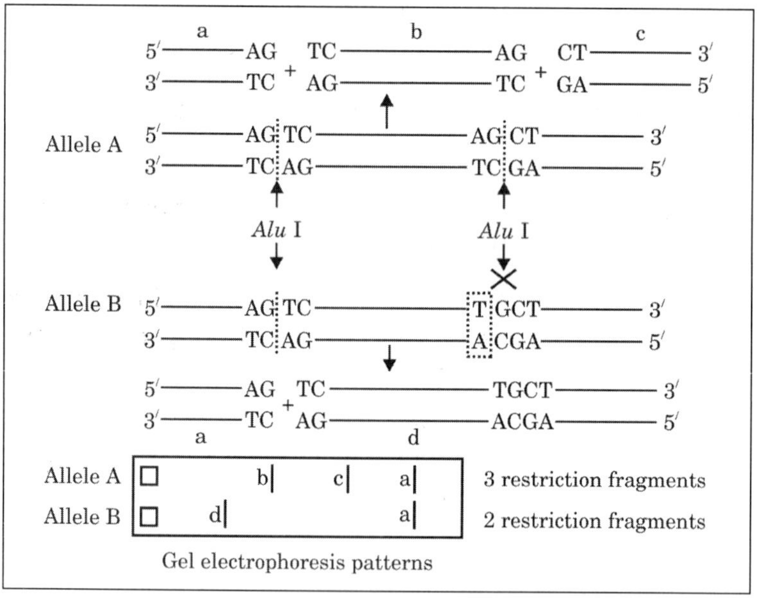

Fig. 7.7: Hypothetical example of restriction fragment length polymorphism (RFLP) Allele A has 2 sites for *Alu*I but Allel B has one site due to single base modification in the 2nd site

SSLPs

Simple sequence length polymorphism (Fig. 7.8) is exhibited by the *minisatellites* (VNTRs) and *microsatellites* (STRs). However, microsatellites are preferred because they are more amenable than minisatellites for PCR and their even distribution in the genome. Minisatellites are longer and occur mainly in the telomeric region. Microsatellites are short DNA sections about 6 to 10 nucleotides in length, occurring consecutively in the $5' \rightarrow 3'$ direction in a chain. The number of sequence repeats in a STR set vary and may be upto around 20.

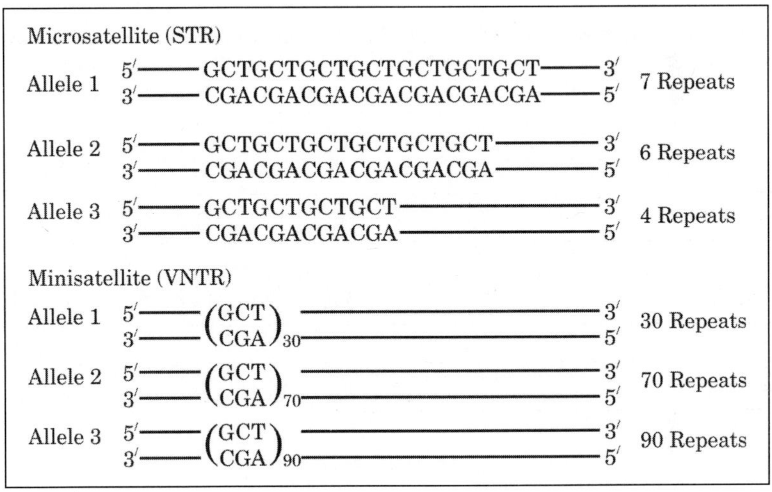

Fig. 7.8: Simple sequence length polymorphism (SSLPs)

Different forms of a STR can again be detected by PCR with primers complementary to the sequences flanking the STR and capillary electrophoresis of the products.

SNPs

These are different forms of a sequence differing in a single nucleotide like ..CAAT***G***TAAC.. and ..CAAT**C**TAAC... SNPs are identified by probe hybridization and several other methods. The probes are made on the basis of preexisting data of known sequences (Fig. 7.9).

The main approach is *oligonucleotide hybridization analysis* and is maximally used in the DNA microarray or DNA chip technology. Synthetic oligonucleotides are synthesized and spotted and immobilized on a glass slide to make a DNA micro array. To make a DNA chip the oligonucleotides are directly synthesized on a wafer of silicon or glass. The sample DNA is fluorescence labeled and applied on the microarray or chip and specific hybrids detected by fluorescence microscopy. As the sequence of the oligonucleotides are known, specific SNPs and their alleles can be detected by this method.

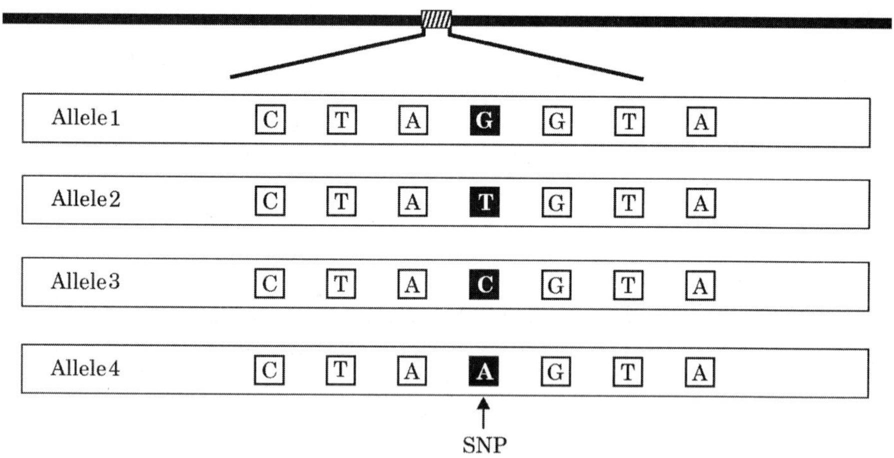

Fig. 7.9: Single nucleotide polymorphism (SNP)

Many variations in responses to drugs in humans have been linked to SN polymorphism. One person may respond to a drug and another may not because of difference in one nucleotide in a sequence set.

The polymorphic forms of any of the above DNA markers are like the alleles of genes. For locating their positions in the genome map, plants or animals with known markers are crossed and DNA from the parents and the resulting offsprings are subjected to linkage analysis as in the case of genes. However, unlike yeast (*Saccharomyces cerevisiae*) and fruitfly (*Drosophila melanogaster*) there are limitations with higher plants and animals of much lower reproductive frequency and hence low numbers of cross overs.

For genetic mapping in humans, where designed crossing is not available, *pedigree analysis* is done. Bacteria also present a challenge. A bacterial cell has a single chromosome, besides plasmids and is thus haploid. In the haploid bacterial cells crossovers cannot take place. Therefore, genetic mapping is done with designed gene transfer during the processes of conjugation, transformation or transduction.

Physical Mapping

Physical mapping is done by molecular biology techniques. As mentioned above, genetic mapping depends on crossover frequency and hence is not very useful in the case of higher eukaryotes. Physical mapping which utilizes direct typing of specific DNA is more accurate and useful.

For physical mapping the popular procedures are, restriction mapping, sequence tagged sites, fluorescence *in situ* hybridization (FISH) and use of mapping reagents.

Restriction mapping

This method is best used with DNA strands of about 50 kb length. For instance, it was used for the restriction mapping of clones of whole genome shotgun fragments. Usually a DNA fragment is subjected to series of degradations by digestions with restriction enzymes. The whole process consists of complete digestion with two restriction enzymes separately, with the two enzymes combined and finally partial digestion with the enzyme for which there are more sites. The fragments are separated in each case and the restriction map deduced (Fig. 7.10).

For larger DNA strands, rare cutting restriction enzymes as *Not* I is used. As the resulting fragments are large, normal fixed field elctrophoresis cannot separate the fragments properly. Therefore, modified agarose gel electrophoretic methods are used. They are orthogonal field alternation gel electrophoresis (OFAGE) and pulse field gel electrophoresis (PFGE). In OFAGE, instead of a pair of

Enzymes	*Fragment number*	*Fragment sizes, kb*	*Total, kb*
*Xba*I	2	24.0, 24.5	48.5
*Xho*I	2	15.0, 33.5	48.5
*Xba*I + *Xho*I	3	09.0, 15.0, 24.5	48.5

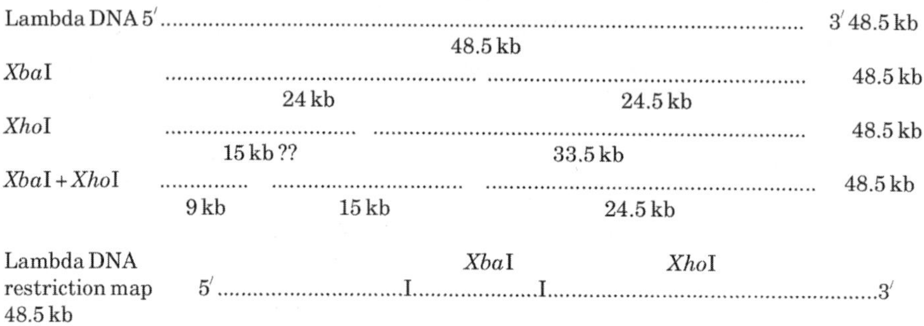

Fig. 7.10: Restriction mapping with example of lambda phage DNA

electrodes at the two poles there are two pairs of electrodes at 45° to the gel plate providing alternate electrical pulses and in PFGE there are additional electrodes at angles besides the main pulse (Fig. 7.11).

A convenient method is optical mapping of gel stretched DNA strands. For this chromosomal DNA is suspended in molten agar solution containing a restriction enzyme and plated on slides. Stretching of the fibre takes place as the gel solidifies. Addition of magensium ions to the gel activates restriction activity. After due incubation the cut sites are visualized by fluorescence microscopy.

Two more types of polymorphic markers are very useful, specially in the case of plants. They are *randomly amplified polymorphic DNA* (RAPD) and *amplified fragment length polymorphism* (AFLP). In both cases prior sequence information is not necessary. RAPD is detected by using specific synthetic oligonucleotides as single primers for PCR. The oligonucleotide primer should be around 10 nuceotides long, with 50 to 80% G+C content and nonpalindromic. Genomic DNA is amplified with this single primer by PCR and the fragments detected by agrose gel electrophoresis. For AFLP, the genomic DNA

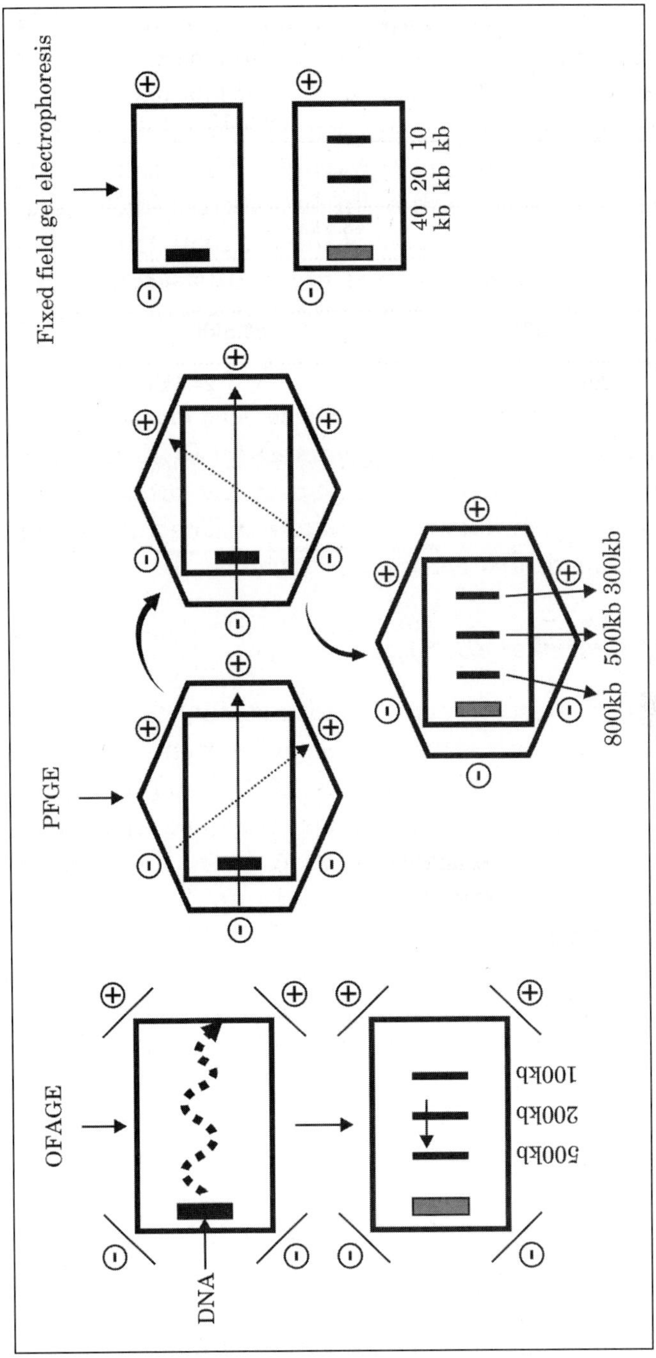

Fig. 7.11: Fixed field electrophoresis, OFAGE and PFGE

is cut with two restriction enzymes, usually EcoRI and MseI, the fragments are separated by poly acrylamide gel electrophoresis. To the ends of the fragments, *Eco*RI and *Mse*I site adaptors are added to act as templates for universal primers complementary to these sites. Amplification is done in two stages. First, the existing digest fragments are amplified with a single selective nucleotide primer. The second amplification is done of the products formed in the first step, using two to three selective nucleotide primers. AFLP has enabled making of high density physical maps of several plant genomes.

The *microsatellites* or STRs can also be used for physical mapping. As mentioned above, they are detected by PCR of a clone with synthetic primers followed by analysis of the products, mainly by polyacrylamide gel capillary electrophoresis. The microsatellites can be sequence tagged also as described below.

Sequence tagged sites (STS)

Sequence tagged sites are some unique sequences in the genome popularized by the work of scientists as Botstein, Olsen and others scientists. These are short sequences, 200 to 500 b long and occur only once in a genome. They are found from the sequences of a genome under study from the existing database. From the sequence data, primers are made, the intervening strand amplified by PCR and used to detect its occurrence in a DNA clone. The BAC clones for human genome sequencing were extensively analysed for STS content. It was found that different types of STSs occur in a clone. Comparison of STS contents of different clones help in mapping. Clones with nearly similar STS content would be overlapping. The sequences could be of genetically mapped microsatellites, some random genomic sequences or built from cDNAs. STSs from c-DNA are termed as *expressed sequence tags* (EST). As the sequences, the STS markers exist in the data bases, any STS can be detected by hybridization with synthetic complementary labeled probe or probes (mapping reagent). However, the preferred method is using PCR with synthetic primers. STS mapping can be done on individual chromosomes or chromosome fragments of a chromosome obtained by flow cytometry or through radiation hybrids.

For flow cytometric separation, dividing cells are ruptured, chromosomes stained with a fluorescent dye and the mix passed through a narrow aperture. Through the aperture, single chromosomes pass, one at a time. The single isolated chromosome is deflected away by a pointed electrical pulse and collected for further studies (Fig. 7.12).

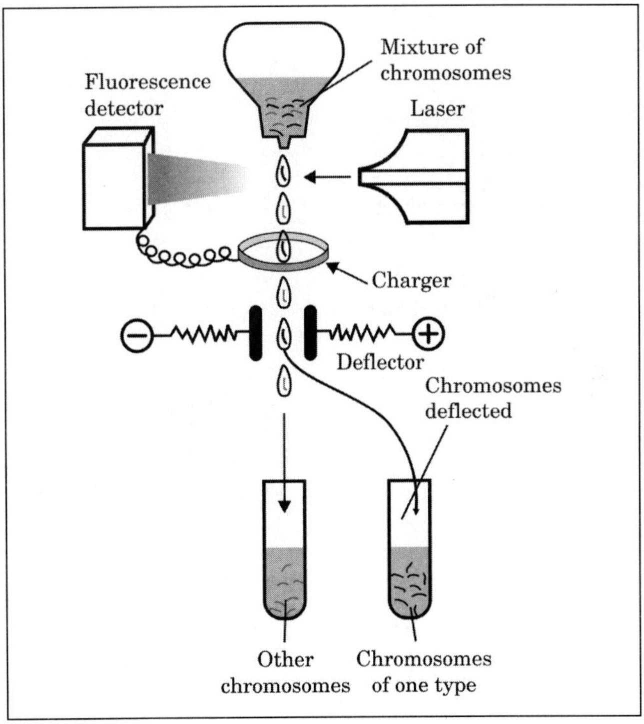

Fig. 7.12: Flow cytometric separation of chromosomes

FISH

In *fluorescence in situ hybridization* protocol, a specific DNA sequence is visualized in an intact chromosome by hybridization with a fluorescent probe. A sequenced cloned DNA is labeled with a fluorescence marker. This DNA probe is used for hybridization with a preparation of chromosmes set on a slide. The probe hybridizes with the complementary sequence in the chromosome and can be visualized under a microscope. Repetition with other clones of the genomic library under study, reveal the positions of the DNA fragments on the chromosome (Fig. 7.13).

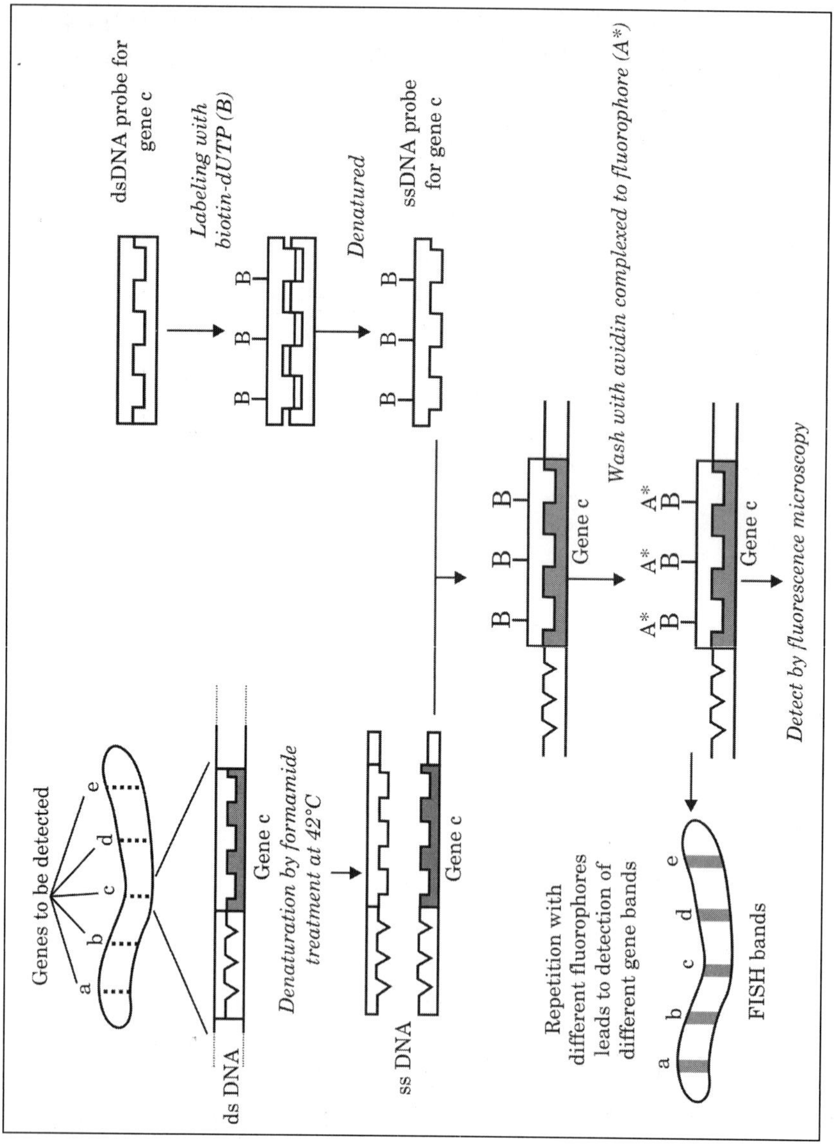

Fig. 7.13: Simplified fluorescence *in situ* hybridization protocol

As several fluorescence probes are available, different DNA strands can be labeled with different probes and separate bands visualized simultaneously. There are two aspects of importance. The DNA in the chromosome is double stranded and has to be transformed into single stranded form for hybridization. For this, after fixation on the slide, the chromosome is treated with formamide which denatures and transforms the DNA fibre into the single stranded form. A problem with eukaryotes is the occurrence of similar repeat units accross the genome. The probe being cloned DNA from the same organism will also have repeat units besides the particular identifying sequence. It can therefore, hybridize at some other site also. To avoid this, the probe DNA is incubated with a solution of unlabeled DNA enriched with repeat units. This treatment blocks the repetitive sequences and keeps only the unique section exposed, for hybridizing with the correct complementary site. The metaphase chromosomes being highly condensed, the bands are around 1Mb apart. The chromosomes may be stretched to some extent without losing identity by centrifugation, bringing the interband distance to about 300 kb.

Somatic cell hybridization with radiation hybrids

Radiation hybrids constitute a mix of hamster cell lines each containing a part of a chromosome. The cell lines are plated on a suitable matrix and bathed with different labeled DNA probes. Hybridization reveals the positions of different sequences on the chromosomes. This method was extensively used in the Human genome project. In the 1960s, G. Barski and co-workers developed the technique of somatic cell hybridization in animal systems. In fertilization, fusion of two haploid gametes leads to the formation of a diploid zygote. In somatic cell hybridization, two diploid somatic cells are made to fuse to form initially a binucleate *heterokaryon* and finally a uninucleate *synkaryon*. The fusion is facilitated by the use of UV inactivated Sendai virus or PEG (poly ethylene glycol). Such somatic cell hybrids can be *monospecific*, formed from fusion of cells of the same species or *interspecific* arising due to the fusion of cells of two different species. In human chromosome mapping, mouse-human cell hybrids have been very useful. They are made by fusing mouse cell lines with established human fibrocyte or leukocyte cell lines. In these hybrid cell lines, the mouse and human chromosomes can be distinguished and both mouse and human genes are expressed. But, the biggest advantage is that, after fusion, during subsequent divisions of the hybrid cells, while the mouse chromosomes are retained, most of the human chromosomes are degraded, leaving back few after around 30 divisions. The degradation of human chromosomes is random and therefore, in different hybrid cells, different human chromosomes are retained. Such cell lines with different chromosomes are used for genetic mapping of human chromosomes (Fig. 7.14). The location of some genes in a chromosome is found by screening for the trait in different cell lines with known human chromosomes or chromosome fragments.

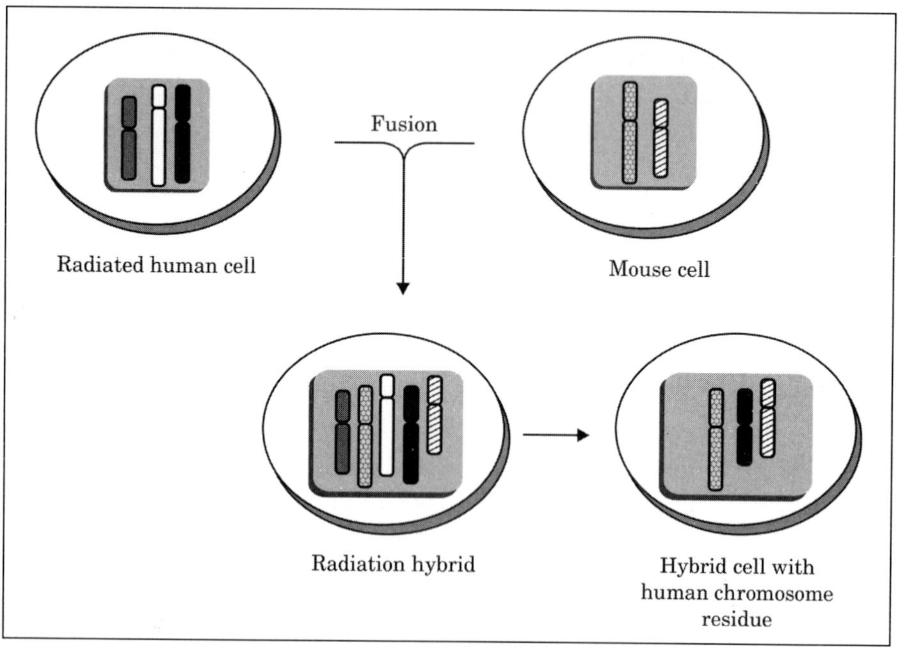

Fig. 7.14: Radiation hybrid cell with one human chromosome residue and all mouse chromosomes

In practice, the hybrid cells are distinguished from parent cells by using a selection medium the HAT medium. This consists of Hypoxanthine, Aminopterin and Thymidine. Aminopterin is a nucleotide biosynthesis inhibitor and both hypoxanthine and thymidine are nucleotide precursors. In the presence of aminopterin, nucleotide biosynthesis is halted and the cells die. However, if in the same medium hypoxanthine and thymidine are added, the cells survive as these precursors are converted to the necessary nucleotides by an alternate salvage pathway. The enzyme *Hypoxanthine-guanine phosphoribosyl transferase* (HGPRT) converts hypoxanthtine to guanine, and *thymidine kinase* (TK) phosphorylates thymidine. The strategy for selection is to hybridize TK^- $HGPRT^+$ mouse cells with TK^+ $HGPRT^-$ human cells in the HAT medium. In this medium only the mouse $HGPRT^+$-human TK^+ hybrid cells survive. Such cells are cultured for several generations in a normal medium and then analysed for traits. During repeated cell divisions many human chromosomes are lost. Clones of individual cells are grown and the chromosomes retained are identified. Such cell lines are tested for known phenotypes of genes for their localization on specific chromosomes. For instance, cells were picked from different clones and grown separately in HAT medium. Out of all the clones only one survived. This surviving clone had chromosome 17. This indicated that the TK gene is present in chromosome 17. More specific localization could be done by sophisticated staining techniques as G-banding and sequencing techniques. The methods of genetic and physical mapping mentioned above helped in making detailed chromosome maps which helped as reference points in whole genome sequencing.

7.5. ASSEMBLING SEQUENCE READS INTO CONTIGS

Approaches for arranging of sequenced clones or *sequence reads* in a linear contiguous order or *contig* have evolved over time. Some of the common methods which were used have been, chromosome walking, clone finger printing by for instance RFLP patterns, repetitive DNA PCR and use of sequence tag sites (STS) as briefly described below.

Chromosome Walking

In *chromosome walking* the genomic DNA is broken down into larger fragments and these are cloned by using high capacity vectors as BAC. Each fragment is sequenced. Thereafter, one of the clones is labeled and used as a probe on the library array. This probe hybridizes with fragment/s having overlapping sequences. For the next round an aliquot of the hybridized fragment is labeled and used again as a probe to observe further overlapping. This process is repeated and from the linear chain of overlapping sequences, the total sequence of clones, the clone contig is deduced. A more rapid method is to use paired end reads as probes for detecting overlaps.

Restriction Fragment Length Patterns

Restriction fragment length patterns are used for fingerprinting of different BAC clones or clones of other high capacity vectors. The clone is cut with a restriction enzyme as *Hind*III and the fragments separated by agarose gel electrophoresis. The pattern acts as a fingerprint for the clone. Fragments with similar patterns have overlapping sequences and form a contig.

For fingerprinting with *repetitive DNA PCR,* the genomic DNA is fragmented and the fragments are cloned with some high capacity vector. Sequencing of such cloned DNA strands obtained after fragmentation of an eukaryotic genomic DNA reveal many repetitive sequences interpersed along the genome sequence. Advantage is taken of this phenomenon to determine overlaps amongst sections of DNA inbetween similar repeats. Therefore, primers are made of a repetitive sequence, labeled and used as probe for the library. The hybridized clones are gel electrophoresed. Labeled bands of similar size indicate overlapping. This process is repeated with other repetitive sequences and the clone contig assembled.

Sequence Tagged Site (STS)

Sequence tagged site (STS) based mapping has been mentioned above. STSs, which can be a gene or a section of nongenic DNA, also helps in revealing overlaps. Labeled PCR primers are made of the STS from available database and used as probes with the genomic library. The clones which hybridize with the STS probes and have similar STS content are overlapping clones. Such data help in assembling a clone contig.

7.6. SEQUENCING OF THE WHOLE GENOME

Sequencing of very long whole genomes had presented many challenges. The DNA sequencing methods, as the chain termination protocol and even automated sequencers can at the most determine sequences of strands of upto 800 bp length. Therefore, for finding the whole genome sequence, the genome has to be fragmented, the sequence of each fragment determined and the *sequence reads*, so obtained have to be reassembled into the correct sequence from $5'$ to $3'$ end. *Accurate sequencing is enabled by the application of some or all of the techniques of genome mapping and clone contig assembly as described above.* Successful sequencing of many microbial, animal and plant genomes could be done because of further advances in DNA science technologies. *These advances mainly include, high capacity cloning vectors* (Chapter 5)*, various DNA markers mentioned above, automated sequencing using capillary gel electrophoresis and computer based assembly algorithms.*

As mentioned earlier, sequencing of prokaryotic genomes with mainly unique sequences is simpler than eukaryotic genomes which have both unique and repetitive sequences. The prokaryotic genomes were constructed mainly by a simple *shotgun sequencing* method for which, availability of chromosome maps was not a prerequisite. Eukaryotic genomes required more evolved methods. Two methods were used initially for human genome sequencing. These were an improved *Whole genome shotgun* method and the *Hierarchical shotgun sequencing* or *Clone contig* method. These or protocols adapted from these approaches have been used for whole genome sequencing of other eukaryotes.

Whole Genome Shotgun Sequencing

Prokaryotes

This method is useful for short prokaryotic genomes and was used for sequencing the *Haemophilus influenzae* genome. For sequencing of this bacterium, the total DNA was extracted and subjected to shotgun fragmentation by sonication. The fragments were separated by agarose gel electrophoresis and the fragment cloned with a suitable plasmid vector to construct a genomic library. The fragments of about 1.5 kb length, were selected and sequenced by chain termination or pyrosequencing. This meant determining thousands of sequences giving proper coverage. These sequence data were fed into a computer database. A suitable algorithm assembled the sequenced cloned fragments, the *sequence reads* with overlaps into *contigs.* Therefore, each contig is a row of contiguous overlapping clones. The software further assembled the contigs into nonoverlapping rows. Thus, sequences of thousands of fragments were narrowed down to few hundred contigs. As some fragments were left out from the clone library, gaps remained in the contig series. Examination of the left out clones revealed some sequences with their end sequences matching the ends of the contigs on two sides of a gap. This strategy closed some of the gaps. Some gaps still remained because some fragments could not be cloned possibly by the cloning strategy used. So, the genome fragments were cloned again with another vector as bacteriophage λ. A microarray of this library was hybridized, one at a time, with oligonucleotides representing the end sequences of the contigs adjoining the gaps. As contig sequences were known, such oligonucleotides could be easily synthesized. Such hybridization revealed matching fragments for the closure of the remaining gaps. The gap sequences were further confirmed by preparing PCR primers of contig end sequences and filling the gaps, as far as possible, with the PCR products formed. From such studies the total sequence was deduced.

This method can have problems with eukaryotic genomes which contain massive amounts of repetitive DNA. If similar sequences are repeated in the genomic DNA then the overlaps obtained may be misleading.

Eukaryotic (human genome)

For human genome sequencing again, the chromosomal DNA was directly sheared without referring to a map by Venter and colleagues at Celera Genomics but, strategies were developed to overcome the problems presented by genome wide repeats in sequencing the big eukaryotic Human Genome. One method was to use fragments longer than the average length of a stretch of repetitive DNA, so that the repeat unit section comes sufficiently within the fragment, far from the ends. Thus 10 kb, 20 kb and 50 kb fragments were preferably used. The whole genome was fragmented by sonication, strands separated, cloned and sequenced. The sequence reads were assembled by a computer algorithm into *contigs* of overlapping sequences. Contigs with small sequence gaps were arranged in a linear order to form *scaffolds*. The scaffolds were also arranged in a row with physical gaps between them (Fig. 7.15). As the sequence of the DNA fragments constituting the contigs are known, the sequence gaps lie between known end sequences or *paired end reads* (mate pairs) of adjoining fragments. Such gaps could be filled by PCR with the paired end reads as primers. However, physical gaps originated in the sequence setting due to lack of clones. These gaps were closed by using STS markers with reference to the physical map of the chromosome concerned, to locate and align the scaffolds in a proper linear order. Through these efforts Celera Genomics arrived at a *draft* sequence of the human genome.

Hierarchical Shotgun Sequencing

The public consortium IHGSC (International Human Genome Sequencing Consortium) centres around the world coordinated by Francis Collins (NIH, US), Aristides Patrinos (DOE, US) and Michael Morgan (The Wellcome Trust, UK) took a tedious but systematic approach to human genome sequencing.

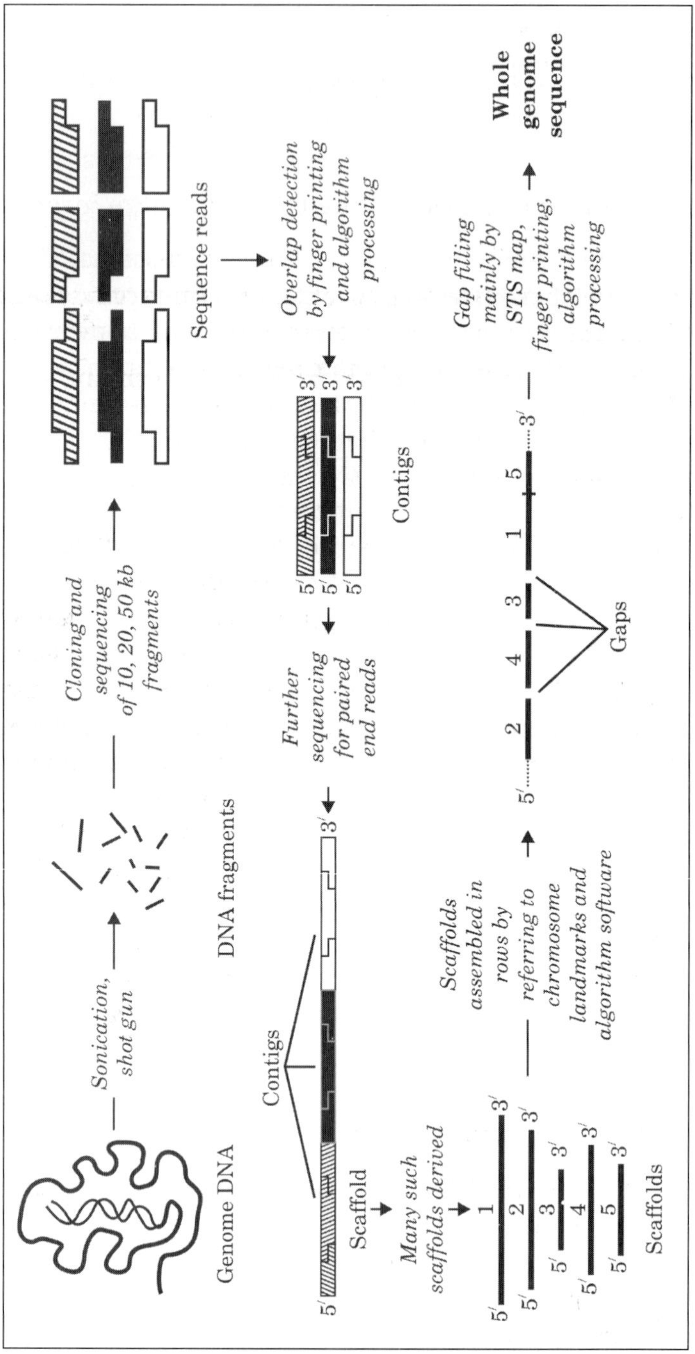

Fig. 7.15: Brief outline of whole genome sequencing by Celera Genomics

In brief, the steps were:

1. Making and refining genetic and physical map of each chromosome to be used as reference points,
2. Extraction of whole or known fragment of chromosomal DNA,
3. Restriction fragmentation of the DNA into large inserts,
4. BAC cloning of each insert,
5. Sequencing of each clone to obtain sequenced strands or *sequence reads*,
6. Determining clone fingerprints by RFL pattern and also STS content,
7. Computer assembly of the sequence reads into a linear order with minimum overlaps, like laying a single row of tiles on a path, that is, generating a *minimal tiling path* of a chromosome region,
8. For gap filling, fragmentation and subcloning of each clone of the tiling path,
9. Sequencing the ends of each insert to obtain *paired end* data or *paired end* reads,
10. Repeating the above steps with all the euchromatic regions of the chromosome,
11. Computer assembly of the total whole genome sequence.

This method was more tedious but systematic, having been based on updated genetic and physical maps of each chromosome.

Basically chromosomal DNA was sheared with rare cutting restriction enzymes as *Lmu*I, *Pvu*I to release large fragments of about 200 to 500 kb size. The fragments were amplified by BAC cloning and the clones were fingerprinted and sequenced. The data generated by these studies were processed for computer based algorithm analysis to generate a clone contig of minimal overlapping clones or a *minimal tiling path* along the chromosome being sequenced. Clones from the tiling path were further restricted and strands of about 3 kb length further subcloned in plasmids and subjected to sequencing of their ends. The *paired end* or the *mate pair data* were assembled by a computer software developed by David Haussler and James Kent to generate a maximum gap filled sequence (Fig. 7.16). The IHGSC relased this *draft* sequence as the 'Golden Path Sequence' in 2000.

So, both groups, the public IHGSC and the private Celera genomics announced their draft sequences of the human genome in June 2000. The formal public announcement was made in a press conference by the then US President William Clinton and Prime Minister of UK Tony Blair.

The two sequences were still unfinished. The IHGC sequence had 7.5X coverage and contained about 150,000 gaps. The Celera Genomics had two drafts WGA (pure whole genome assembly) of 5.1X coverage without reference to mapping data and CSA (compartmentalized sequence assembly) made with the aid of IHGSC mapping data and having a combined coverage of 12.6X. CSA sequence had approximately 170,000 gaps. However, these data revealed many important features of the human genome including the fact that the total number of genes is around 30,000, much lower than the previously estimated number of 100,000. Further, location of many disease related genes helped in investigations on gene therapy.

As mentioned above, the draft sequences contained many gaps. These gaps were obviously in the individual large BAC inserts. Enunciation of a complete sequence needed filling of the gaps in each insert obtained from each chromosomal DNA and finally the whole genome. It may be recalled that in the IHGSC project, the sequence of each of these large inserts was obtained, by assembling the sequences of 3 kb subclones derived through its fragmentation. These gaps could, therefore, be because of three main reasons.

1. No DNA fragments matching the end sequences of strands on either side of a gap, available in the subclone library.
2. Some such matching DNA fragments were available in the subclones but missed out.

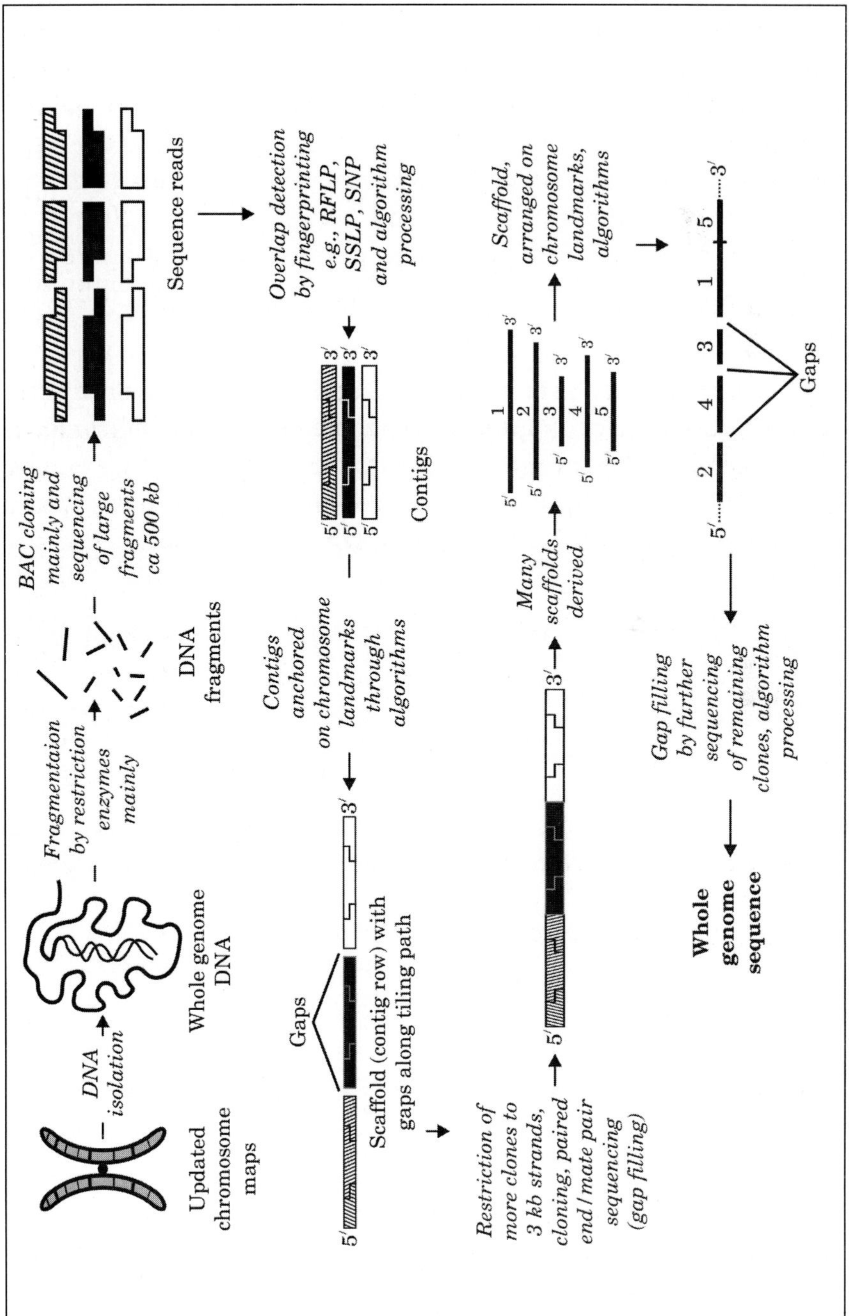

Fig. 7.16: Brief outline of hierarchical genome sequencing by the Public Consortium (IHGSC)

3. The base pair quality was poor for assembling.

The first gap was mainly filled by making the intervening sequence with PCR using the sequence ends on either side of the gap as primers. The second type of gap was filled by finding overlaps through hybridization of some of the available subclones with end sequences on two sides of the gap and between them, referred to as *primer walking*. The third type of gap had been filled mainly by resequencing the fragments on either side of the gap with different DNA polymerases.

The *finished* sequence of the IHGSC endeavour, now containing only 281 gaps instead of 150,000 and covering 99.3% of euchromatin was submitted in the GenBank. The formal announcement was made on 25th April, 2003, the Golden Jubilee anniversary date of the publication of the double helix model of DNA, by Watson and Crick in Nature.

A major feature of the HGP was inclusion of parallel sequencing of other genomes as of yeast *Saccharomyces cerevisiae*, fruit fly *Drosophila melanogaster*, and a nematode *Caenorhabditis elegans*. Sequencing of some plant genomes as of *Arabidopsis thaliana, Oryza sativa* and *Zea mays* were also initiated. Rapid advances in sequencing equipment, use of robotics and advanced softwares have given rise to fast Next Generation genome sequencing techniques.

The technologies mentioned above came into being because of the onset of initiatives for the Human Genome Project in mid 1980s. A brief recap of the milestones is given in section 7.8.

7.7. SOME PLANT GENOMES SEQUENCED / BEING SEQUENCED

To date fifty five plant genomes have been sequenced or are being sequenced. Examples of some common plants are:

Arabidopsis lyrata, 2011 (model plant), *Arabidopsis thaliana* 2000 (model plant), *Oryza sativa* ssp *indica* 2002 (cereal), *Oryza sativa* ssp *japonica* 2002 (ceeal), *Triticum aestivum*, 2012 (cereal), *Zea mays*, ssp *mays* B73 2009 (cereal), *Jatropha curcas*, Palawan 2010 (biodiesel source), *Solanum tuberosum* 2011 (potato), *Musa acuminata* (banana) 2012, *Hordeum vulgare* (cereal), *Solanum lycopersicum* (tomato) 2012, *Cajanus cajan* (Pigeon pea) 2012, *Glycine max* (soybean) 2010. The details can be seen online (*e.g.*, http://genomevolution.org/wiki/index.php/Sequenced_plant_genomes).

7.8. SOME MILESTONES IN THE HISTORY OF HUMAN GENOME SEQUENCING

The idea of HGP was first given in a meeting including eminent scientists as Botstein and Gibert in 1985 at the University of California, Santa Cruz. The meeting was conducted by the then Chancellor of UCLA, Santa Cruz, Robert Sinsheimer. He had worked on ÕX174 and was inspired by Professor Sanger's sequencing of its genome. Initially, many doubts were expressed as: (a) it is a monotonous technical work and not a creative science, (b) it is very expensive, as at a rate of even $ 1 per base, the minimum budget would be $ 3 billion for the approximately $3{\times}10^9$ long human genome, (c) requires cooperative work of many groups and just not individual efforts, (d) why sequence the whole genome, why not just map and sequence the genes. However, Charles DeLisi, Director, Office of Health and Research, the Department of Energy, US, decided to initiate work on human genome. This was mainly motivated by a paper of Renato Delbucco in *Science*, advocating the value of human genome sequencing for understanding cancer. That shook up the biologists. The National Research Council and many scientists at the National Institute of Health under the stewardship of James Wyngaarden were activated in favour of HGP. Professor James D Watson further emphasised the need for whole human genome sequencing. The project attracted the business sector and Celera Genomics, headed by J. Craig Venter, announced its venture on HG sequencing. Finally, NIH of US, endorsed HGP in 1988 and soon NIH and DOE signed an MOU for a joint effort for human genome sequencing. An added feature was that, HGP included sequencing of other organisms also, including *Escherichia coli, Drosophila* and

Caenorhabditis elegans. Professor Watson became the head of the public endeavour. Later Prof Watson resigned being against patenting of human sequence ESTs and Dr. Francis Collins, took over.

A public consortium of research centres developed, *the G5*. The consortium included Washington University, St Louis, US; Baylor College of Medicine, Houston, US; Whitehead Institute, MIT, Cambridge, US; Joint Genome Institute of DOE, Walnut Hills, US; Fred Sanger Institute, Cambridge, UK (supported by Wellcome Trust).

Fifteen other sequencing centres in US, France, Germany, Japan and China joined the league. Over time, a variety of advances in sequencing technology brought down also the costs of sequencing, from an estimated $ 1 per base to 10 cents per base. At present, due to the efforts of a private venture, Illumina, the sequencing is very rapid and also costs even lesser.

Professor Watson decreed, at the very beginning in 1988, that 3% of the budget be spent on examining the Ethical, Legal and Social Implications (ELSI) of HGP.

The ELSI programme has been an unprecedented effort in examining the interaction of Science and Society.

Another historical landmark was the formulation, in 1996, of the *Bermuda principles,* developed in a conference of scientists sponsored by the Wellcome Trust. The principles directed that all sequence data generated by the sequencing groups be automatically released to public databases within 24 hours of their assembly. These principles made the data *open access* and prevented blanket patenting of small sequence segments having importance in sequence assembly but of no known utility otherwise.

8

Production of Transgenic Plants

SUMMARY OF CONTENTS

Transgenic plant production - Step 1. *Primary Transformant* by genetic engineering, Step 2. *GM crop for release* to farmers by transgene transfer from primary transformant to a crop variety by conventional plant breeding; some marker genes in plant transformation; marker free transgenic crops; promoters; plant tissue culture; plant breeding basics; transgenes and transgenic crops; testing of GM crops - transgene analysis and performance

Plants have been the mainstay for the survival of humans. Plants have provided daily needs as food, medicines, structural items and other useful products. Crop improvement for food security and other livelihood needs has been a human endeavour for over 10000 years. Therefore, attempts for improvement of plants have gone hand in hand with the progress of civilization. The approach has mostly included selection of variants of a species with desired traits, crossing regimes for transfer of genes of interest to developing varieties, production of high yielding hybrids and others. All these conventional practices have two disadvantages: (a) a limited gene pool, restricted within the species and sexually compatible species, as the source of genes, (b) inability to separate and transfer useful genes linked closely to a deleterious gene on a chromosome. As genetic engineering enables gene transfer across species, the gene pool resource becomes much larger extending over a broad spectrum of species. Further, the transgenes are processed in the laboratories as independent units. Therefore, the problems of interference due to linkage do not arise.

Since the first production of transgenic tobacco by Otten *et al.,* in 1981, many transgenic plants have been produced specially with qualities of better resistance to biotic and abiotic stresses and higher nutritional value, at the laboratory level. Similarly, genetic engineers have transformed plants to act as bioreactors or factories for therapeutic proteins and industrial enzymes. These researches have led to the founding of a successful agriculture biotechnology industry.

Out of the methods used for plant transformation, the most commercially used method is *Agrobacterium* mediated transformation, constituting 65% of total production followed by nearly 20% by particle bombardment technology and 15% by other methods.

Transgenic plants are produced in two steps (Fig. 8.1). In the first step, the *primary transformant* is produced by transferring a transgene of interest to a plant by *recombinant DNA technology* or *genetic engineering*, in a laboratory. Transgene insertion / the product is sometimes referred to as 'event'. In the second step, the transgene again is transferred to a short listed agriculturally suitable variety, by *conventional plant breeding.*

Therefore, there are two steps in GM crop production:

Step 1: Genetic Engineering: Production of the Primary Transformant by recombinant DNA technology.

Step 2: Plant Breeding: Transgene transfer to an agriculturally viable crop variety by conventional breeding.

8.1. STEP 1. PRODUCTION OF PRIMARY TRANSFORMANT (BY RECOMBINANT DNA TECHNOLOGY)

Genetic modification of plants is done in two ways, mostly by *gene addition* and in some cases by *gene silencing,* sometimes referred to as *gene subtraction*. For gene addition, two types of protocols are

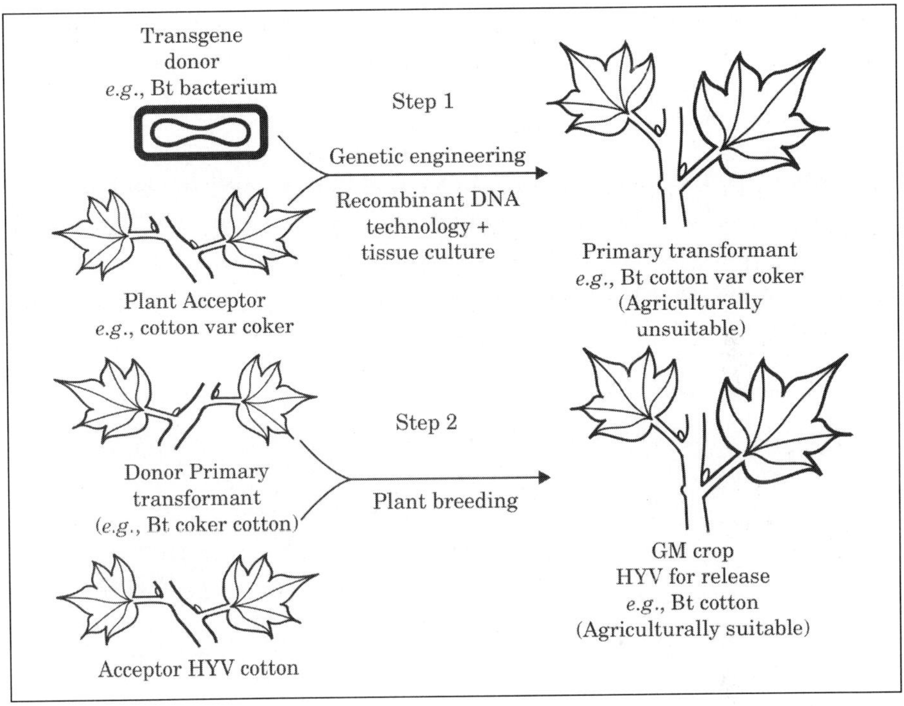

Fig. 8.1: Steps of GM crop production

used for delivery of DNA into plants, the *indirect mediated transfer* (IMT) methods and the *direct DNA transfer* methods. The *Agrobacterium* mediated DNA transfer and direct physical DNA transfer by *particle delivery system* (PDS) are most common protocols. The DNA transferred is the transgene of interest alongwith the noncoding regulatory sequences which ensure its proper expression. Plant transformation by gene silencing is done mostly by some RNA types as, antisense RNA, Si RNA and micro RNA.

Gene Addition by Indirect, Mediated Transfer

Agrobacterium tumefaciens mediated gene transfer

The protocol simulates the processes of bacterial DNA integration into host plant DNA during *Agrobacterium* infection. It would, therefore, be of interest to know how this natural genetic engineer works.

A. tumefaciens the natural genetic engineer

Agrobacterium tumefaciens is a gram negative soil bacterium. On coming in contact with wounded plant stem tissue, usually at the soil plant interface, it induces tumour formation. As the portion of stem at the soil plant interface is called 'crown', the tumour is referred to as *Crown Gall Tumour* (Fig 8.2).

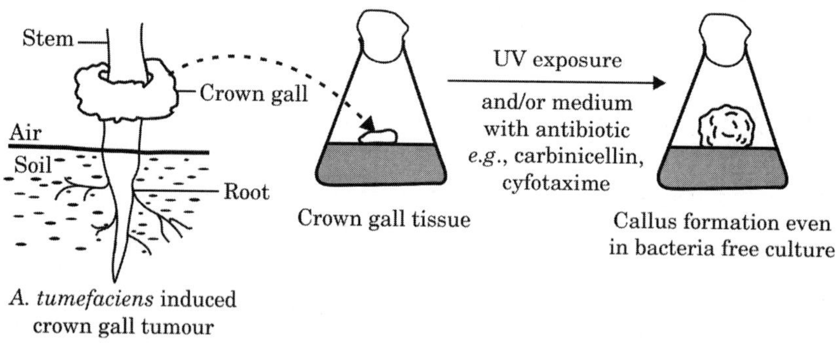

Fig. 8.2: *Agrobacterium tumefaciens* induced crown gall tumour formation

Bacterial origin of this crown gall formation was established nearly a century back. The wounded tissue releases compounds as sugars, phenols and specially acetosyringine which attract and activate the bacteria. It is interesting that after infection and callus formation, the bacterium is not needed for further cell divisions and growth of the callus. Excised callus tissue, made bacteria free, by treatment with UV light and antibiotics, as carbenicellin, and put on a basal tissue culture medium grows vigorously even without added plant growth regulators. Thus, the transformation is truly *oncogenic* in nature. This was a puzzle for a long time, up to the late sixties. Intensive molecular biological studies involving comparison of the DNA of the bacterium, healthy tobacco and of crown gall tissue had been conducted to get an answer. The studies revealed that the bacterium induces tumour formation by transferring a portion of its DNA, the T-DNA, to the host cell followed by integration of this T-DNA fragment with the host DNA. Further, detailed studies indicated the nature, steps and the mechanism of the DNA transfer which triggers tumour formation.

The bacterium *Agrobacterium tumefaciens,* like many other bacteria, contain some extra-chromosomal circular DNA, the plasmids. The circular plasmid DNAs of bacteria help in conjugation with other bacteria and transfer of traits to the conjugated bacteria. The plasmids of tumour inducing virulent strains of *Agrobacterium* are referred to as Ti plasmids. Molecular genetic studies have indicated the sequence and components of these plasmids (Fig. 8.3). There are two types of Ti plasmids, the nopaline and octopine type, the former driving nopaline synthesis and the latter opaline. The main components of the Ti plasmid circular DNA are, the *T DNA* region which is transferred and induces unbridled cell divisions or *oncogeny* and synthesis of opines; a sequence for opine catabolism; a sequence needed for *conjugation* to other cells; sequences needed for infection or virulence, the *vir* region, and a sequence needed for starting replication, the origin of replication ori. The T DNA is flanked by 25 bp characteristic repeats, the *left border* (LB) and the *right border* (RB) sequences. There is an *over drive sequence* near the RB in octopine type Ti plasmids (Fig. 8.3).

Analysis of the T- DNA has shown it to contain several genes as those for plant growth regulators. There are genes for synthesis of the phytohormones, auxins and cytokinins referred to as the *oncogenes* and also the *opine synthesis* genes for the synthesis of some nitrogenous compounds, the opines. Opines are unusual amino acid derivatives, as octopine and nopaline, not found in untransformed plants. On integration of the T-DNA to the nuclear DNA of the host plant cells, its oncogenes synthesize the phytohormones in the host cells leading to uncontrolled cell divisions and tumour formation. At the same time the opine synthesis genes start the synthesis of opines. The opines ooze out of the tumours, are washed down and provide nutrition for the *Agrobacterium* residing in the intercellular spaces of the tumour tissue and the soil below. These

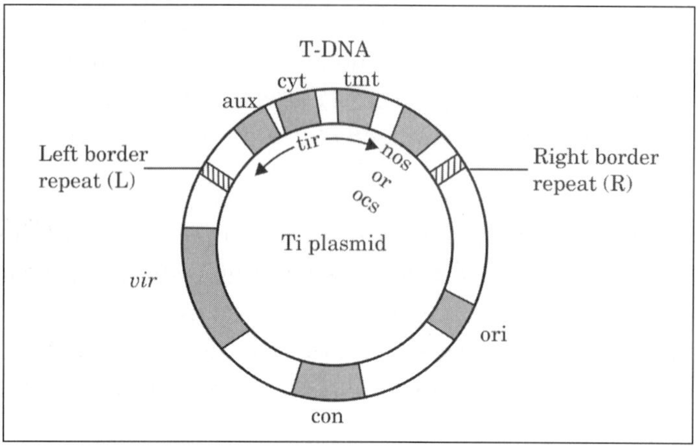

Fig. 8.3: Generalized Ti plasmid structure (aux = auxin gene, cyt = cytokinin gene, tmt = tumour size regulator, tir = tumour inducing region, nos = nopaline synthase, ocs = octopine synthase, *vir* = virulence gene, con = conjugation sequence, ori = origin of replication)

strains of *Agrobacterium* are therefore, *natural genetic engineers*. They transform the host tissue they come in contact with, to vigorously growing tumours actively manufacturing nutrients for the bacteria (Fig. 8.4).

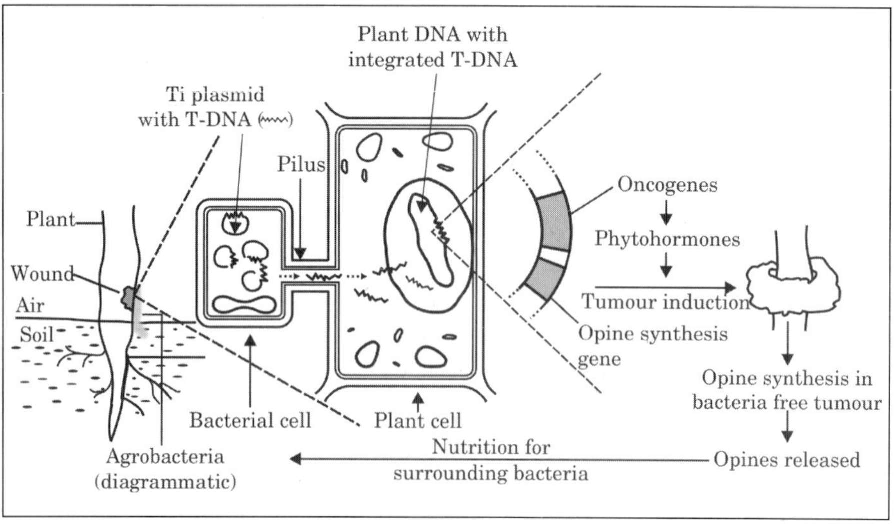

Fig. 8.4: *Agrobacterium*, the natural genetic engineer

[After bacteria-plant cell connection, T-DNA moves to plant cell and integrates with plant nuclear DNA. Phytohormones encoding genes (oncogenes) synthesize hormones and induced tumour. Opines synthesized in tumours on a massive scale provide nutrition for the surrounding bacteria]

Understanding the details of this process has led to the designing of vectors and protocols for *Agrobacterium tumefaciens* mediated transformation of plants. The border 25 bp sequences are necessary for the transfer of T-DNA or any other DNA fragment to be inserted in a plant host genome. Further, infection and transfer of T-DNA cannot take place in the absence of the *vir* genes. There are six main *vir* genes. They are referred to as *vir* A, B, C, D, E and G and occur on the plasmid in the sequence ...A...B...G...C...D...E.... Studies have indicated that these genes carry out specific functions in the transfer process through proteins encoded by them. The proteins encoded by the *vir* A, B, G, C, D and E genes are termed as VirA, VirB, VirG, VirC, VirD and VirE, respectively. It may be noted that *vir*A is strongly constitutive, that is, it is transcribed spontaneously and continuously without stimulation from any inducer molecule. The *vir*G is constitutive but the activity increases in the presence of wounded plant extracts. Both are monocistronic and encode single protein products. The genes *vir* B, C, D, E are plant inducible, that is, are activated by some chemicals produced by the wound tissue of the plants. These genes, in contrast to the *vir*A and *vir*G genes, are polycistronic, producing several proteins from the same mRNA.

The *vir*A gene product VirA is present on the inner bacterial membrane and being a chemoreceptor perceives the presence of some phenolic compounds known as acetosyringine, released by the wounded plant tissue. Signal transduction from VirA, a kinase, transphosphorylates VirG protein leading to its activation. The activated VirG, a transcription factor, stimulates the expression of other *vir*B, C, D and E genes by binding to their operator sequences. The *vir*D gene produces VirD proteins. VirD1 is an endonuclease and it causes a nick on the bottom strand of the T-DNA, at a specific site between the third and the fourth base of each 25 bp repeat.

This action of VirD1 is enhanced by VirC proteins. Another T-DNA strand is synthesized in the $5' \rightarrow 3'$ direction, on the exposed bottom strand starting from the nick on the right border. VirD2 protein and the products of *vir*E gene, the VirE proteins, seem to stabilize and protect the T-DNA during its transport through the plant cell cytoplasm to the nucleus and final integration into the host chromosomal DNA. The VirB proteins get located on the bacterial envelope and form a pilus, a channel between the bacterial cell and the plant host cell, for the T-DNA to move across to the host cell (Fig. 8.5). There are some genes on the bacterial chromosome also

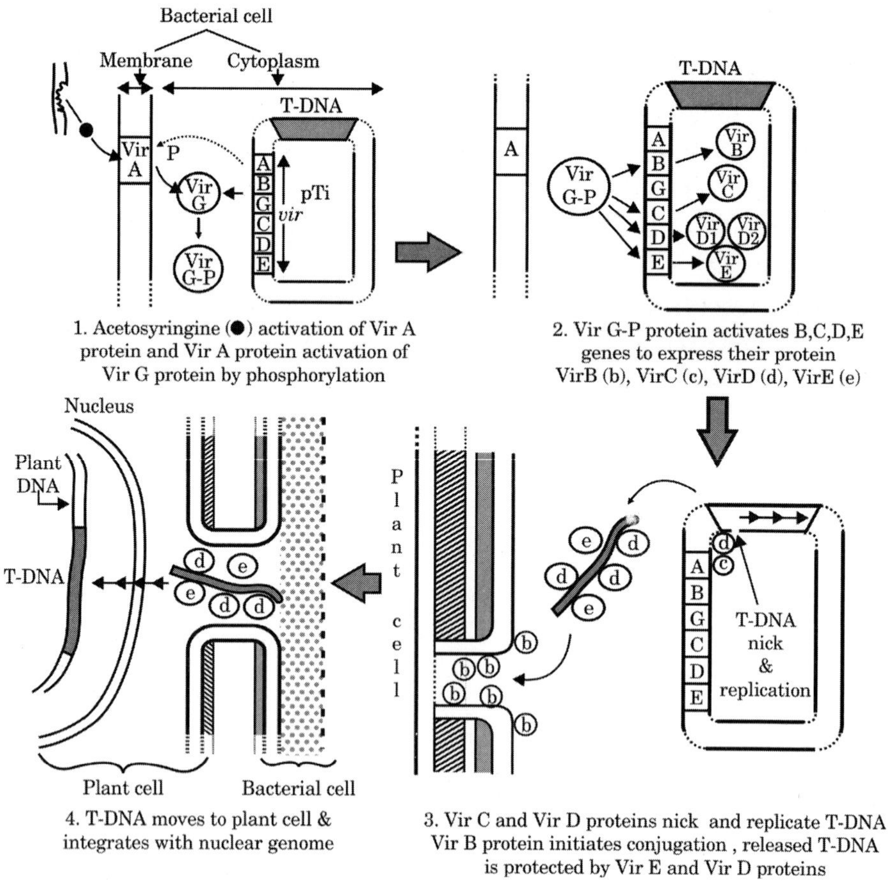

Fig. 8.5: Concerted action of the *vir* genes for T-DNA transfer

which help in this T-DNA transfer process. Further, the nopaline and octopine Ti plasmids differ slightly, the nopaline plasmids being simpler.

The realization that, for T DNA transfer, the sequences needed in the Ti plasmid are the right and left 25 bp repeat borders and the *vir* genes and that the other *sequences in between the L and R borders are not needed* led to the designing of useful Ti plasmid based vectors for gene transfer to plants. The Ti plasmid could be disarmed by removing the segment between LB and RB and replaced with a desired transgene for transfer to a target organism.

The following components are necessary for Ti plasmid derived vectors:

(a) Two separate *origins of replication* for multiplication of the vector-transgene recombinant in *E. coli* and also *Agrobacterium tumefaciens.*

(b) A *multiple cloning site* for insertion of the transgene of interest.

(c) A bacterial *selectable / screenable* marker as also a plant *selectable / screenable* marker.

(d) *Promoter* sequences for markers and the transgene inserted.

(e) A set of genes that ensure in the *Agrobacterium* cells, the scission of the transgene cassette from the vector-transgene recombinant DNA and its safe passage to the host plant genome.

This function is carried out by the *vir* genes already present in the Ti plasmids.

Figure 8.6 gives a comparative idea of the Ti plasmid and a hypothetical disarmed vector.

The above list indicates that the Ti plasmid derived vectors require much more sequences than normal vectors. Incorporation of all these sequences in the Ti plasmid of around 200 kb size would lead to a large vector molecule. Such large constructs are not suitable for transformation protocols. Therefore, with much ingenuity, two different types of Ti vectors have been designed, the *cointegrate* and the *binary* vectors. There is an interesting difference between the two types of vectors. In *cointegrate*, the *vir* genes are near the transgene in the same vector complex and therefore can act on the neighbouring transgene, that is can act in a *cis* mode. In the binary vector system, the *vir* genes although present on a different vector, in the cell, can still induce transfer of the transgene, that is can luckily act from a distance in the *trans* mode.

It may be mentioned, that now the cointegrate vectors are of only historical importance. For current transformation protocols, only binary vectors are used. However, the innovation in cointegrate vector designing is of interest and is briefly recapitulated here.

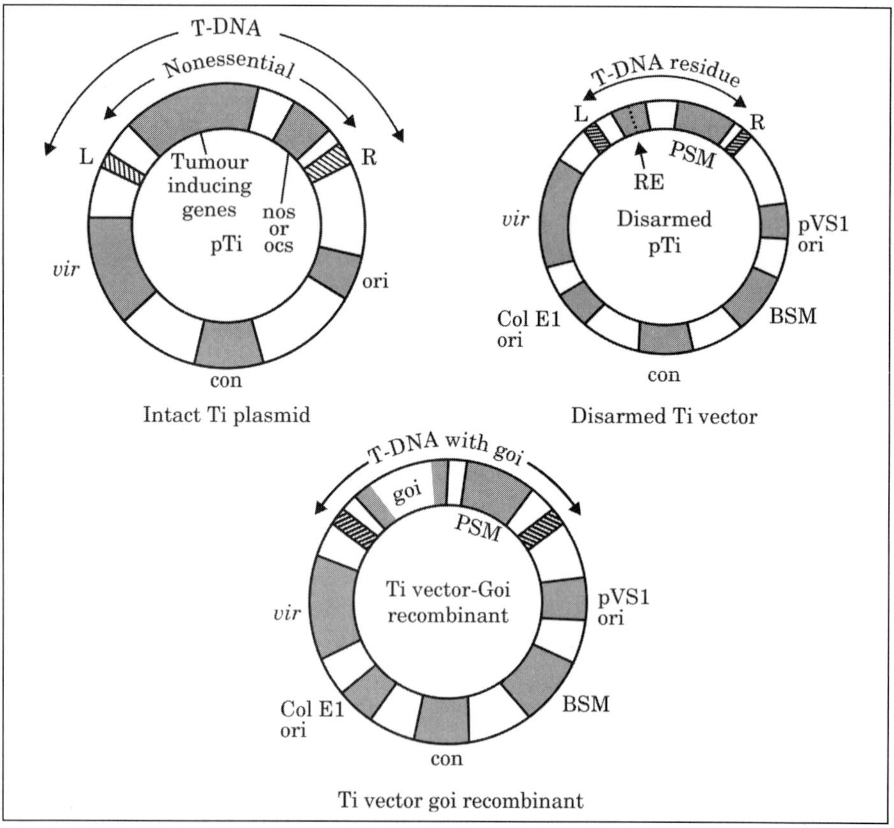

Fig. 8.6: Generalized Ti plasmid, disarmed vector and recombinant

For construction of cointegrate vectors, two vectors, one containing the gene insert and the other with the *vir* genes were made to combine, that is get integrated into each other. Nearly similar T DNA sequences in each of the two, enabled them to integrate by homologous recombination. In one system, a nopaline type Ti plasmid vector was disarmed by removal of most of T-DNA between left and right borders and insertion of a pBR322 vector sequence (pGV3850). This vector was introduced in *Agrobacterium* cells. At the same time, the gene of interest was inserted in a small *E. coli* intermediate vector, also derived from pBR322 (pGV1103). This intermediate vector was first multiplied in *E. coli* and then through conjugation transferred to pGV3850 vector containing *Agrobacterium*. In the *Agrobacterium* cells the two vectors integrated by homologous recombination between the pBR322 common sites in the T DNA region, result in the formation of the pGV3850::pGV1103 cointegrate vector. The final construct resembled the hypothetical vector–GOI recombinant shown in Fig. 8.6. This system had the disadvantage of being too large and hence of low efficiency.

Binary vectors

In this vector system, advantage is taken of the fact that *vir* genes can act in the *trans* mode also. The *vir* genes of one vector molecule can enable the transfer of T-DNA present in another vector molecule of the same cell. A typical binary vector is pBin19 (Fig. 8.7).

In this vector, again the oncogenes are removed between left and right borders. The main sequences from left border onwards are, a *lacZ'* sequence (coding for α-peptide of β-galactosidase) with a mutiple cloning site of several restriction recognition sequences and a plant selectable kanamycin resistance gene (neomycin phosphoryl transferase II, nptII with nopaline synthase promoter, nos) near the right border. Besides these are the prokaryotic kanamycin resistance gene (APH-I), which works both in *E. coli* and *Agrobacterium* and the RK2 derived origin of replication and conjugation sequences. This vector is about 12 kb long. The transgene of interest alongwith requisite promoter as CaMV 35S promoter and a terminating code,

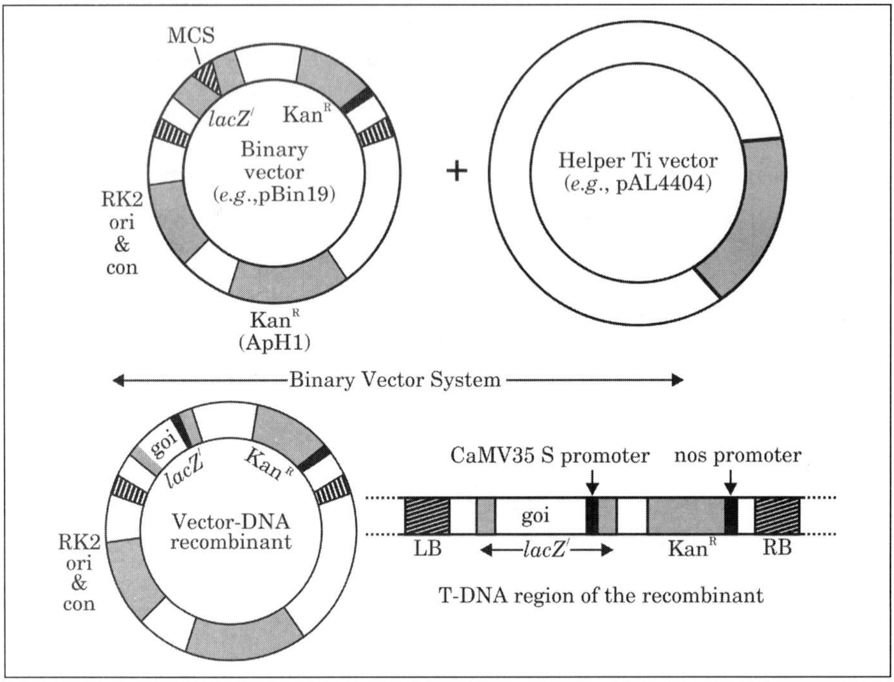

Fig. 8.7: The binary vector system for *Agrobacterium* mediated plant transformation (goi=gene of interest, rest as in previous figures)

is inserted in the *mcs* region. This transgene-vector recombinant DNA is maintained in *E. coli.* As the transgene insertion leads to inactivation of the *lacZ'* gene, the recombinants are recognized by lack of blue colour development on X-Gal-IPTG plates (Chapter 5).

A common Ti helper II plasmid vector, as pAL4404, is made by removing the T DNA and mainly retaining the *vir* genes. These are around 20 times larger than the main vector because of the presence of the *vir* genes and are maintained in the *Agrobacterium* cells. The transgene-pBIN19 recombinant DNA is transferred after purification and testing, to the pAL4404 containing *Agrobacterium* cells usually by electroporation. Such *Agrobacterium* cells containing both components of the binary vector system are further used for plant tranformation. It may be noted that in several *Agrobacterium* strains, the resident T. plasmid can act as a helper plasmid for the pBIN19 binary vector.

The pBIN19 vector is relatively large being about 12 kb in size. Further, the T DNA containing the transgene enters into the host plant cell from the right border end. Therefore, the plant selectable marker enters before the transgene. If somehow the entry gets interrupted the marker will reach but not the transgene. Thus, the resulting products will give positive indication despite the missing transgene, a false indication.

Better vectors as pGreen have been developed. pGreen vector is smaller, being about 4.6 kb in size and with the plant selectable marker near the left border (Fig. 8.8). In this case the transgene enters before the marker and a positive test means a true indication.

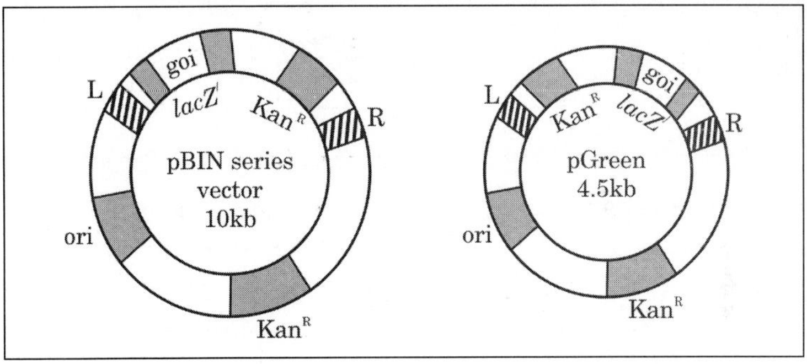

Fig. 8.8: Advances in binary vector desiging (pBIN type is larger and can give false positive selection as the selection marker reaches before goi with the right border. pGreen is smaller and ensures that selection marker enters after goi)

Another related bacterium, *Agrobacterium rhizogenes* induces hairy root condition in plants via the Ri plasmid. However, this plasmid has had limited application in plant transformation.

A. tumefaciens mediated plant transformation

The most commonly used vector system is the *Binary* vector system, pBIN19 and pAL4404 combination. The steps of a common protocol are shown in Fig. 8.9. The transgene of interest is inserted in the multiple cloning site of the binary vector pBIN19. This vector carrying the transgene is introduced and maintained in *E. coli*. The helper

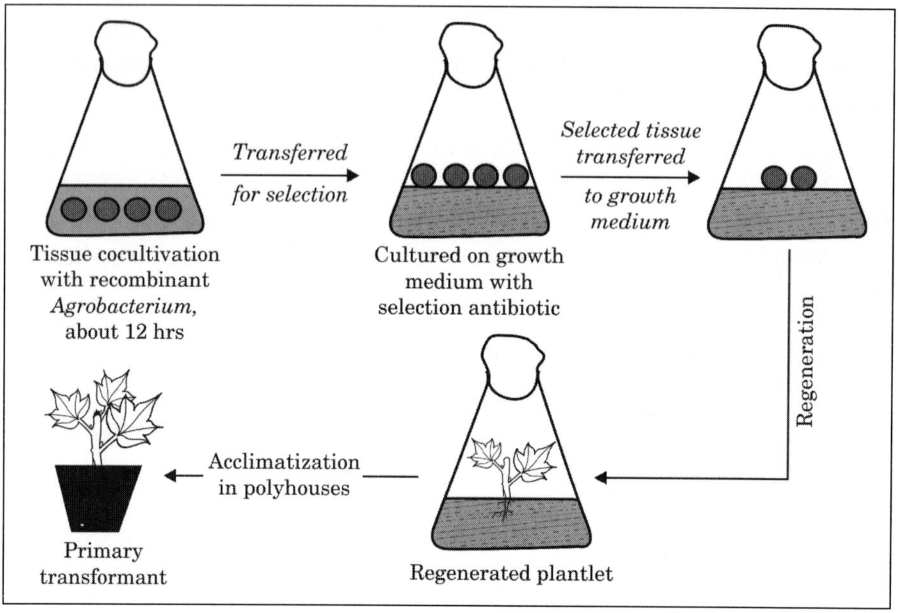

Fig. 8.9: Basic steps of *Agrobacterium tumefaciens* mediated plant transformation

plasmid pAL4404 with *vir* genes is kept in a suitable strain of *Agrobacterium.* For conducting transformation, the pBIN19 vector is transferred by electroporation to the helper plasmid containing *Agrobacterium*. Such *Agrobacterium* cells with the binary vector system are *cocultivated* with the target plant cells or selected growing tissues. After the required incubation period the transformed tissues are washed with an antibiotic solution as carbinicellin or cefotaxime solution, which kill *Agrobacterium* cells but do not affect the plant cells. Thereafter, the plant tissue explants are processed for selection. The selection protocol depends on the marker genes inserted alongwith the transgene. If the marker is the commonly used kanamycin resistance gene, the *Agrobacterium* cocultured materials are transferred to an agar medium containing kanamycin. The transformed cells or tissues with the transgene can only survive.

The selected tissues are grown to plantlets and plants via proper tissue culture techniques as described in a later section.

Lot of research is on developing other microbial transfer systems specially with non parasitic symbiotic bacteria as *Rhizobium* and related species.

Electroporation

In this method (Fig. 8.10), DNA strands, circular or linear, are made to enter cells through electrical pulses. Short electrical pulses induce reversible permeability in the cell membranes, possibly through pore formation. However, the method requires preparation of protoplasts from plant tissues. This is achieved by treating the tissue with a mixture of cell wall degrading enzymes. The protoplasts to be transformed are incubated with the transgene-vector DNA dissolved in a suitable buffer medium with required additives. The suspension containing protoplasts and DNA is given electrical pulses in the range of 100 to 300 secs at around 120v/mm. This makes the DNA to migrate through the widened cell membrane pores into the cells and further to integrate with chromosomal DNA.

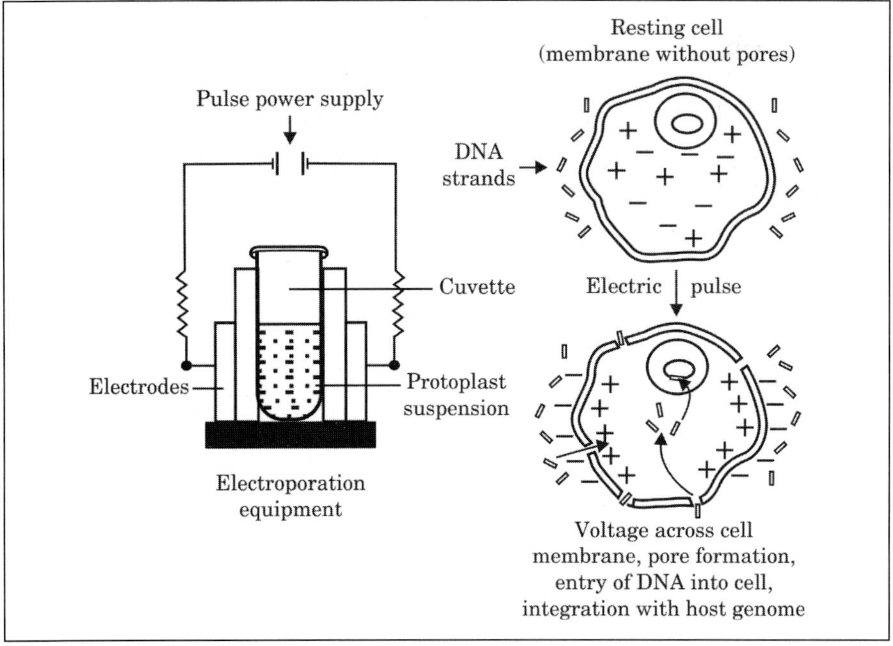

Fig. 8.10: Electroporation equipment and principle

Addition of compounds as spermidine in the incubation medium and condensed single strandedness of DNA help in integration. The electroporation process is less damaging than particle bombardment. Many crops, specially cereals, have been transformed by electroporation.

There are also other methods like *chemical mediated* DNA delivery to plant cells. For instance, polyethylene glycol (PEG) or polyethylene alcohol have been used for transfers into cells suspended in high pH, Ca^{2+} containing media. These methods however, have limited applications.

Gene Addition by Direct DNA Transfer

Particle bombardment

This is a widely used method and is carried out using particle bombardment devices. Various synonymous terms are used for this technique like, *biolistics, microprojectile bombardment, gene gun* and *Particle Delivery System* (PDS) method.

The DNA fragment of interest is precipitated on the surface of gold or tungsten particles and these *microcarriers* are plated as a monolayer on a Mylar sheet membrane, a *macrocarrier.*

The device used is a vertical cylinder having from top downwards, a high pressure helium chamber sealed by a disc at the bottom, a platform to place the macrocarrier, a protective mesh screen and a chamber for the plant tissue (Fig. 8.11).

The whole setup simulates a gun barrel except that the target is right at the end of the barrel, always at point blank range. The macrocarrier is inserted below the helium box and the equipment triggerred. Release of helium at high pressure acts like ignited gunpowder and propels the macrocarrier towards the screen. The macrocarrier hits the mesh screen and the force of the impact, shoots the DNA coated heavy metal particles into the cells or tissue. After

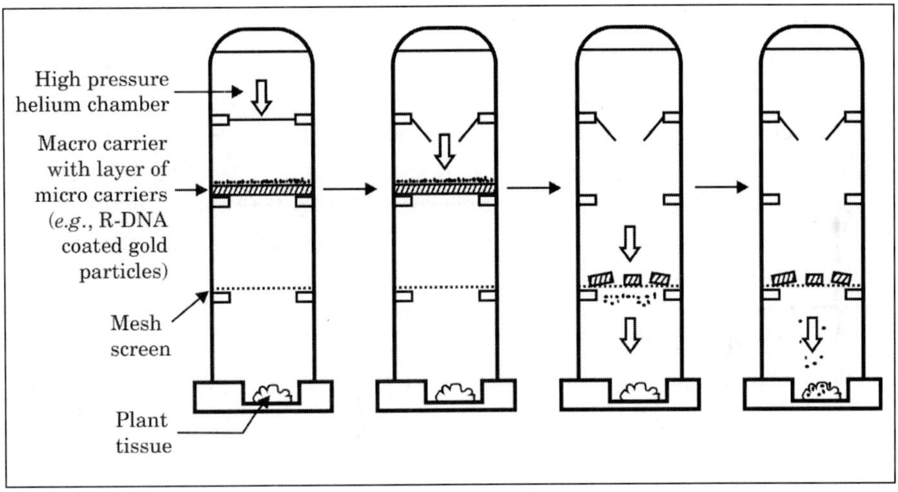

Fig 8.11: Particle delivery system (biolistics, gene gun, microprojectile)

penetration, the DNA detaches from the particles and some integrate with the plant cellular genome. A partial vacuum in the casing in which the target tissue is kept helps in better uptake of DNA by the cells.

Various preparatory optimization efforts ensure successful transformation. The heavy metal particles are washed in ethanol and distilled water and mixed with transgene-DNA-vector in a solution of $CaCl_2$, some polyamine or aminosiloxane. After binding, the mix is centrifuged, the pellet washed in ethanol and suspended in ethanol. This suspension is layered on a membrane and dried to make the macrocarrier. The choice of polyamine or related compound is important. Usually, the transgene-DNA is spliced into a vector with the marker genes. Sometimes the marker gene, a selective agent as a gene for antibiotic resistance and a reporter gene are kept on separate vectors (the co-transformation method). This helps in removal of the marker from transformed plants at a later stage through segregation in the progeny plants. The target tissue is normally a regenerating embryonic tissue and is kept in a slightly hypertonic medium. Slight plasmolytic condition seems to help in the uptake of DNA. Care is taken to keep the force of bombardment such that the penetration number remains below 5 particles per cell. A day after bombardment the calli are processed for selection. From selected transformed explants, plantlets are grown, acclimatized in growth chambers, transferred to soil and further grown to maturity. This technique was specially suitable for production of GM cereal crops. However, *Agrobacterium* mediated methods for cereals have also been developed. The gene gun method is the most preferred method for chloroplast genome engineering.

The use of silicon carbide fibres and microinjection for direct DNA delivery to plant cells has also been practised with some success.

Gene Silencing (Subtraction)

In this method, an existing gene is silenced for better performance of a desired trait. The first commercially released GM crop Flavr Savr tomato was made this way. By gene silencing, the ripening and undesirable softening of tomato could be delayed for sometime. This prevented rotting of tomato during transit from farms to distant markets and ensured the delivery of ripe turgid tomatoes. At the onset of ripening, the enzyme polygalacturonidase is produced. This enzyme degrades the polygalacturonic acid component of cell walls leading to softening of tomato fruit tissues. If the transit time exceeds the time span of ripening, the tomatoes get over ripe and unmarketable, causing huge losses to the grower. Thus, some delay in ripening helps. For this, a portion of the polygalacturonidase gene is synthesized and cloned into a vector in the reverse direction (Fig. 8.12). In tomatoes transformed with this reversed polygalacturonidase DNA an antisense mRNA is produced. This antisense RNA binds with the normal polgalacturonidase mRNA, thus preventing its further translation into the enzyme protein. In this way, in the transformed tomato plants, through silencing of the polygalacturonidase gene, ripening is slowed down.

The transformed cells or tissues are regenerated and grown by the tissue culture protocols (described in a later section) to produce the Primary Transformant.

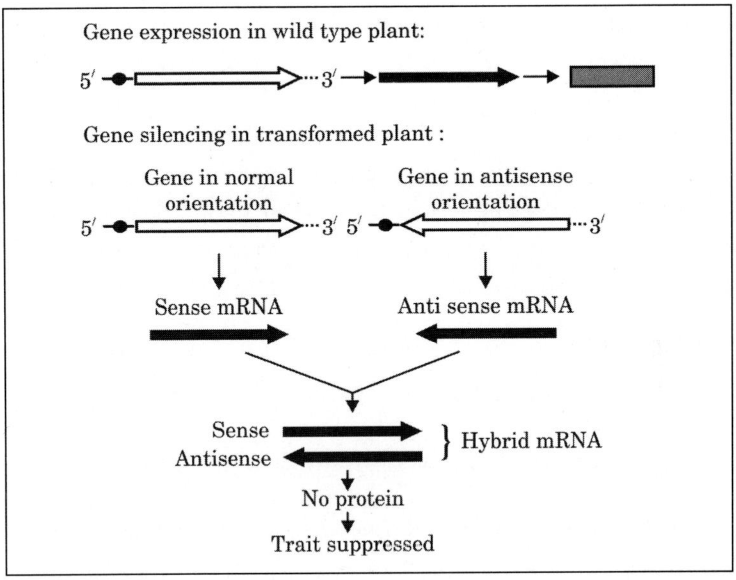

Fig. 8.12: Transformation by gene silencing with antisense RNA

8.2. STEP 2. TRANSGENE TRANSFER FROM PRIMARY TRANSFORMANT TO A CROP VARIETY/HYBRID BY CONVENTIONAL PLANT BREEDING (THE FINAL GM CROP, READY FOR RELEASE TO FARMERS)

What is not mentioned properly in most texts is this second crucial step in GM crop production. The primary transformant is not agriculturally suitable. For some inherent reasons, cells of many highly developed varieties are not competent enough for DNA uptake and therefore, the initial introduction is done in varieties which are amenable for *transformation* and *regeneration*. Some such recipients may even be equivalent to improved land races. For instance, the primary event in Bt cotton is a cultivar Coker (*Gossypium hirsutum* var Coker). Bt cotton-Coker is not agriculturally viable and hence the Bt gene is transferred from it to agriculturally suitable varieties of cotton. For such transfers, the probable varieties are subjected to trials in different agroclimatic zones and suitable ones are shortlisted for the transfer processes. The preferred plant breeding protocol is transgene transfer to a selected variety through backcrossing (Fig. 8.13). This method is briefly described in the following section.

Bt-primary transformant-A (low yield)	×	Non Bt high yielding variety HYV-B

↓

F_1 seeds collected and plants grown

↓

Seeds of high yielding progeny with Bt trait of A selected

↓

Back Cross 1: Plants grown and crossed back with HYV-B

↓

Seeds of high yielding progeny with Bt trait of A selected

↓

Back Cross 2: Plants grown and crossed back with HYV-B

↓

Back crosses repeated till about 90% progeny are high yielding like HYV-B and have Bt trait (takes 5-6 generaions)

↓

Seeds of high yielding plants like HYV-B and with Bt-trait selected

↓

Progeny of individual seeds grown

↓

Again, seeds of HYV-Bt trait plants selected and the process repeated at least for two generations

↓

Homozygous Bt-HYV, GM crop

Fig. 8.13: Simplified back cross protocol for selection of a dominant gene insert with example of transfer of Bt toxin gene from primary transformant to a high yielding variety

8.3. SOME MARKER GENES IN PLANT TRANSFORMATION

Successful and biosafe transformation of plants depends on vectors, markers and promoters. Scientists constantly strive to bring in improvements in these elements for greater efficiency and safety. Vectors have been dealt with in this and previous chapters. In the following sections, some commonly used marker genes have been described.

The efficiency of transformation in both prokaryotic and eukaryotic cells is low. Therefore, it is necessary to include some marker gene/s in the vector-transgene recombinant DNA which will help to identify and distinguish transformed cells from the untransformed ones. The marker genes confer some unique trait in the transformed cells which help in their identification. As indicated in the previous chapters, such traits are of different types as, antibiotic resistance, herbicide resistance, emission of fluorescence upon irradiation and others.

There are two broad groups of markers, a) the *selectable markers*, as, the kanamycin resistance gene and b) *screenable markers* or reporter genes, as, Green fluorescent protein.

Some commonly used marker genes are described in the Tables below (Tables 8.1 and 8.2).

Table 8.1: Selectable markers in plant transformation

Gene	***Abbre-viation***	***Source***	***Selection agent / Parameter***
Antibiotic resistance			
Neomycin phosphotransferaseII	*npII*	*Escherichia coli*	Kanamycin
Hygromycin phosphotransferase	*hpt*	*E. coli*	Hygromycin
Bleomycin resistance	*ble*	*E. coli*	Bleomycin
Herbicide resistance			
Phosphinothricin Acetyl transferase	*bar*	*Streptomyces hygroscopicus*	Phosphinothricin
5-enolpyruvateshikimate 3-phosphatesynthase	*aroA*	*Agrobacterium tumefaciens*	Glyphosate
Alternate carbon source / Growth			
Mannose-6-phosphate isomerase	*manA*	*E. coli*	Mannose
Xylose isomerase	*xylA*	*Streptomyces rubiginosus*	Xylose

Table 8.1: (*Contd...*)

Table 8.1: (*Contd...*)

Gene	*Abbre-viation*	*Source*	*Selection agent / Parameter*
Isopentenyl transferase	*ipt*	*Agrobacterium* sp	Growth without cytokinin in medium

Table 8.2: Screenable markers (reporter genes) in plant transformation

Gene	*Abbreviation*	*Source*	*Detection*
β-Glucuronidase	*gus* / / *uidA*	*E. coli*	Histochemical/ fluorometric
Green fluorescent protein	*gfp*	*Aequorea victoria* (jelly fish)	Fluorescence
Luciferase	*luc*	*Photinus pyralis* (fire fly)	Luminiscence

Selectable marker gene confers resistance against some toxic chemical to the transformed tissues or some other attribute for their selection. As mentioned in previous chapters the marker genes are introduced in the vector DNA in appropriate positions. A common widely used marker is neomycin phosphotransferase II (*nptII*) which encodes a phosphotransferase which destroys the antibiotic kanamycin. Kanamycin resistance markers work best in dicotyledonous plants. For monocotyledonous plants, bleomycin or hygromycin resistance have been found to be better. The *hpt* gene encodes the enzyme hygromycin phosphotransferase which destroys hygromycin. The *bar* gene codes for the enzyme phosphoinothricin acetyl transferase which detoxifies the herbicide phosphoinothricin. All these markers are obtained from bacteria and hence for their expression in plant cells it is necessary to use proper promoters. Commonly used promoters include the *nos* (nopaline synthase of *Agrobacterium*) gene promoter and CaMV35S promoter. Best is to replace their original bacterial promoter and other regulatory elements with those of plant origin. Such chimeric genes with bacterial coding region and plant regulatory sequences enable better recognition of the gene by the plant transcriptional and translational machinery for successful expression. As the antibiotic and herbicide resistance genes used in GM crops may constitute environmental risks, other selection markers have also been isolated from microbes, developed and tried. One example of such biosafe selection markers are those which contribute the trait of survival on alternate carbohydrate sources such as mannose, xylose and other nutrients to the transformed tissue. Thus, transformed tissues with the *manA* gene have an isomerase which converts mannose to glucose, enabling its entry into the glycolytic cycle. For selection, such tissues have to be grown on agar media containing mannose and no antibiotic is involved. Again, for proper functioning in plant cells, plant regulatory elements have to be spliced into the vector upstream and downstream of the marker gene. Another selection method without antibiotics utilizes the *ipt* gene. This gene through isopentenyl transferase expression enhances cytokinin production. Transformed callus tissues with higher proportion of endogenous cytokinin than auxin differentiate shoot tissue only and can grow in cytokinin minus medium. By growing the selected plantlets in media with proper cytokinin:auxin ratio, normal differentiation is restored.

The *screenable reporter* genes have been a great aid in understanding the expression status of the inserted transgenes and therefore, in optimizing transformation protocols. These genes introduce traits which help in visible detection of the transformed tissues. The most commonly used reporter genes are β-glucuronidase (*gus*), green fluorescent protein (*gfp*) and luciferase (*luc*).

The *gus* gene is widely used for both quantitative and qualitative estimates of the inserted genetic elements. For just detection of the occurrence of the gene, that is qualitative assay in transformed tissues, histochemical assay with X-gluc (5-bromo-4-chloro-3-indolyl-β-glucuronide) is done. For quantitative assay, transformed tissues are treated with the fluorogenic substrate MUG (4-methylumbellierryl-β-D-glucuronide) and fluorometric assay of the emission estimated. The *gus* gene is a favourite marker as it is nondestructive and does not require any substrate. Being nondestructive, assays can be conducted on same tissues over time, like continuous monitoring in growing seedlings. Further subcellular visualization of processes can be done by methods as confocal laser scanning microscopy. Activities of genes can also be monitored through linkage with *gfp* gene resulting in fusion protein products. Tissue specific expression of *gfp* has also been achieved using tissue specific promoters. For instance, with CaMV35S promoter, the green fluorescent protein appears in all tissues but, is expressed in pollen only when linked to the pollen specific LAT59 tomato promoter. For functionality in plants some modifications are necessary as, removal of a cryptic intron and making the codons more plant adapted.

The *luc* gene from firefly encodes the luciferase enzyme which catalyzes ATP dependent oxidation of luciferin leading to light emission. Detection of rapid emissions with sensitive recorders as lumimeters, make the *luc* assay useful in detecting low level or localized expression of genes. The sensitive, nondestructive detection also allows for real time assay of expression systems.

8.4. MARKER FREE TRANSGENIC CROPS

Markers are used for selection of the recombinant transformed cells, tissues or plantlets. After transformation and stabilization of the product, the marker genes are no more needed and can be removed. In some cases in which several transformation events are conducted with the same cells or tissues, a selection marker is needed for each gene inserted. If however, a marker gene can be removed before insertion of the second transgene, the repeated transformations can be carried out with the same marker. It may also be necessary to remove some markers, as, antibiotic- or herbicide- resistance markers for which there are biosafety concerns. Therefore, several innovations have been deveolped for removal of markers to obtain marker free transgenic crops. The main methods are cotransformation, use of transposons and also recombinase systems.

Cotransformation

In this technique, the transgene and the marker gene are transferred separately but simultaneously in the same cells or tissues so that, they get preferably, inserted into separate chromosomes. Usually the marker and the transgene are inserted in two separate plasmids. In *Agrobacterium* mediated transformation, the two plasmids may be lodged in different strains or the same strain. Sometimes, a plasmid with two T DNAs, one with marker and the other with the transgene may also be used. As, the two genes occupy different chromosomes, through segregation, marker free transgenic lines can be obtained.

Transposon Mediated Removal

Transposons may also be used for marker removal. These genes encode some enzymes which can cut the neighbouring gene and place it in some other chromosome. In one case the *ipt* marker gene was associated with the *Ac* transposase gene. The ipt gene induced shooty phenotype in the tansformed tissue. Thereafter, the *Ac* transposase gene was induced which caused removal of the *ipt* gene and led to recovery of normal growth pattern in the transformed tissue.

Site Specific Recombinase System

This strategy utilizes the phenomenon of occurrence of a recombinase and its recognition sites in the same organism. Some phages have the Cre recombinase and its specific *lox* site, for example, the Cre/*lox* of bacteriophage P1. Induction of the recombinase results in its cutting the sequence bordered by *lox* sites and removing it or relocating it to another position in the genome with *lox* site. Other examples of recombinase/specific site, paired occurrence are the FLP/*FRT* of yeast *Saccharomyces cerevisiae* and R/*RS* system of *Zygosaccharomyces rouxii.* In one promising strategy the transgene, the selection marker and the recombinase gene are incorporated in the same vector. An inducible promoter is attached upstream of the recombinase gene. The marker and the recombinase gene sets are kept in tandem and bordered by the recombinase recognition sites. After selection of the transformants, the recombinase gene is activated by induction and the recombinase enzyme formed, deletes both the recombinase and the marker genes together.

8.5. PROMOTERS

All genes need promoter sequences to switch them on. Further, every gene requires the promoter and other regulatory elements for proper expression. In a developing organism some genes may have to be on, all the time in all the tissues. Generally however, some genes need to be expressed in certain tissues only and/or at certain times of development. Thus, in nature, there are different types of promoters. Some promoters, the constitutive promoters, constantly and continuously work, that is always keep the genes 'on'. There are also promoters which normally remain inactive and are activated or induced when required by some chemical or other signal. Such promoters can be tissue specific or non specific depending on whether they are functional in only some tissues or in all tissues. Promoters can therefore be *constitutive* or *inducible* and both can be *nonspecific* or *tissue specific.*

The promoters are normally species specific, working in one or related species. In this context, plant transformation has posed many challenges regarding the choice of promoters. The transgenes are transferred across species mainly from bacteria to plants, for instance, the Bt gene from a bacterium to a crop like cotton or maize. Promoters for the bacterial genes may not and usually do not exist in plant genomes. Thus, promoters of microbial origin have to be used and made to work. These promoters are needed for the transgene of interest as also the marker genes.

Intense research by scientists led to the discovery of promoters suitable for the alien genes transferred to plants. The most widely used promoters are the nopaline synthase (*nos*) gene from the Ti plasmid of *Agrobacterium* and CaMV35S promoter from the cauliflower mosaic virus. These are both strong constitutive promoters and nonspecific, being able to work in all tissues. Therefore, they are highly suitable for the expression of selectable marker genes.

These promoters however, work best in dicot plants. For monocot cereals, the useful promoters are *actin* of rice and *ubiquitin I* of maize. There are other promoters also available for expression of alien genes in plants. These are being discovered through *promoter trap* and *enhancer trap* experiments. One such strong constitutive promoter is the tCUP promoter.

The position and orientation of the promoters for the selection gene and the transgene are also important for expression of these genes. If the promoter of one gene is near another, specially near CaMV35S gene, there can be silencing of its gene.

The promoters mentioned above are highly suitable for nuclear tranformation. For chloroplast transformation, the most efficient promoter is the ribosomal RNA operon promoter *Prrn.*

8.6. PLANT TISSUE CULTURE

The transformed cells or tissue explants are subjected to some plant tissue culture protocol for regeneration and development of the whole plant, the primary transformant. *Without regeneration there is no transgenic plant production*. Therefore, a brief account of plant tissue culture is given in the following section.

After the target material (cells, protoplasts, a tissue or organ as leaf) has been transformed it has to be regenerated into whole plants. This is done through tissue culture techniques. The advantage with plants is the inherent property of totipotency of most of the living cells. Theoretically, cells of a plant tissue can dedifferentiate and redifferentiate to form any other type of tissue in a suitable growth medium. Thus, phloem cells in proper growth media can dedifferentiate into undifferentiated callus and redifferentiate to form shoot or root or both under different conditions.

For transformation purposes, cell cultures, protoplast cultures or explants are used. The selected transformed material is transferred aseptically to liquid or solid growth medium (Table 8.3 Murashige and Skoog, MS medium) in glass vessels, as flasks, which are kept in walk-in growth rooms or growth chambers.

Table 8.3: A common tissue culture MS medium

Nutrients	*Concentration, g/l*
Macronutrients	
NH_4NO_3	1.65
KNO_3	1.90
$CaCl_2.2H_2O$	0.44
$MgSO_4.7H_2O$	0.37
KH_2PO_4	0.17
Micronutrients	
KI	0.0008
H_3BO_3	0.0062
$MnSO_4.4H_2O$	0.022
$ZnSO_4.7H_2O$	0.0086
$Na_2MoO_4.2H_2O$	0.0002
$CuSO_4.5H_2O$	0.00002
$CoCl_2.6H_2O$	0.00002
$FeSO_4.7H_2O$	0.027
$Na_2EDTA.2H_2O$	0.037
Organic compounds	
Myoinoistol	0.100
Nicotinic acid	0.0005
Pyridoxine-HCI	0.0005
Thiamine-HCI	0.0005
Glycine	0.002
Sucrose	30

The tissue culture room is maintained at a temperature of about 25°C, illumination of 25 to 50 µmol $m^{-2}s^{-1}$ and sterile conditions. The transformed cells or tissues grow to plantlets via *organogenesis* from calli formed or *somatic embryogenesis* (Fig. 8.14).

The most commonly used *growth medium* is the Murashige and Skoog medium, the MS medium. It consists of soluble macro and micronutrients, EDTA (ethylene diamine tetra acetic acid) chelated

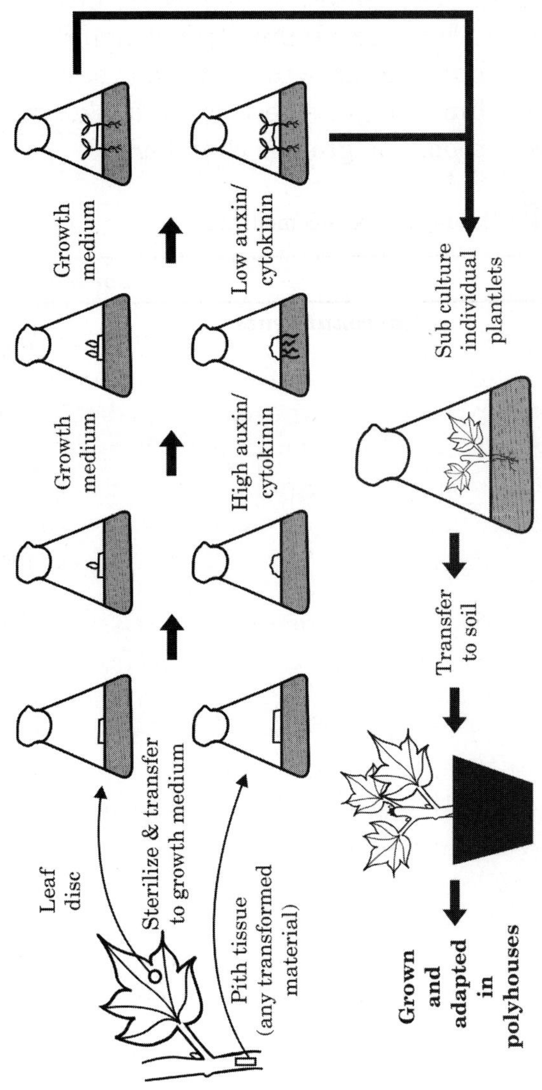

Fig. 8.14: Basic steps of plant regeneration through tissue culture

iron, sucrose as carbon source, growth hormones being mainly auxins and cytokinins, vitamins and other additives as myoinositol and a pH of around 5.5 (Table 8.3). The medium is sterilized before being used and all tissue culture operations are conducted under aseptic conditions.

Callus cultures are made by transferring tissues to media which allow cell division but not cell differentiation. A callus therefore is a mass of undifferentiated cells.

Cell supension cultures are made by breaking down loose friable calli. Such dispersed cells can be maintained as batch cultures for various purposes.

Protoplast cultures are obtained normally from cell cultures. The cell walls are degraded by treatment with cell wall hydrolyzing enzymes as cellulase and pectinase. After the required incubation with the enzymes, the released protoplasts are isolated by sieving through a proper mesh to remove the debris, washed to remove traces of degrading enzymes and floated on sucrose. These are then cultured in an appropriately adjusted (marginally > isotonic) osmotic medium to prevent bursting. Being fragile a better option is to maintain the protoplasts in an alginate matrix and culturing on solid media.

Plantlets from the transformed cells or tissues are developed through *organogenesis* or *somatic embryogenesis*. In the former case, the tissue transformed may form a callus for a brief period and then via phytohormonal regulation differentiate into shoot and/or root or whole plantlet. A higher cytokinin:auxin ratio leads to shoot formation only, lower proportion to roots only and a balanced ratio to plantlets. Direct plantlet development without the callus phase can also be induced. Somatic embryogenesis can also be direct or via a callus stage and is governed by hormonal balances. Like natural embryogenesis pathway, morphogenetic development of the transgenic tissue progresses through globular, heart, torpedo, cotyledonary to seedling stage.

When the seedlings are bigger they are transferred to soil in pots and grown into saplings. The transformed seedlings or events are grown under contained conditions in polyhouses (polythene covered green houses) to select the best performing event. The best event is the one: (a) with high optimal expression of the transgene and (b) is equivalent to the parent in all other aspects. Such selected saplings are acclimatized to natural conditions and allowed to mature for seed formation.

Plants grown from these seeds are the primary transformants or events. From these plants the transgene is transferred to agriculturally profitable varieties by plant breeding methods for commercial release.

8.7. PLANT BREEDING BASICS

It is necessary, even for the scientists involved in recombinant DNA science to be acquainted with the essentials of plant breeding methods. This would help in regulating the proper choice of breeding technique being used for transfer of the transgene from the primary event to an agriculturally suitable variety.

Crop improvement has been a major endeavour since the beginning of agriculture around 10,000 years ago. This was part of the overall endeavour of *domestication*, bringing wild species of plants and animals under human management. In the earlier phases, *selection* played a major role. Selection of some edible versions of wild type crops, vegetables and fruits, having evolved through natural selection, must have been the starting initiatives. This led to further efforts as crossing of crops. The *landraces*, with favourable traits were used as parents, crossed and progeny with a combination of good parental traits selected. This ushered in the processes of *facilitated fertilization* followed by *artificial selection* in plant breeding protocols. Most plant breeding programmes for crop improvement contain elements of crossing and selection and lead to production of *varieties* with reusable seeds. A major discovery was the phenomenon of hybrid vigour or *heterosis* in crossing experiments. In several cases, a trait or some traits in the progeny of a cross can be of higher level than either parent. This is the phenomenon of *heterosis* which introduced *hybrids* in agriculture programmes. In the present times, the Indian farmers use two types of crops, the *varieties* (*desi jati*, in Hindi) with reusable seeds and *hybrids* (*sankar jati,* in Hindi), the seeds of which can be used only once as, in the next generation there is segregation. The main aim of all plant breeding programmes is to produce improved crops with a combination of stably heritable useful traits. Self pollinated and cross pollinated (out crossed) species are bred partly differently.

Self Pollinated Crops

There are several methods. Two useful practices are the single seed bulk breeding and back cross.

Single seed bulk breeding

In this method two homozygous parent lines are crossed, seeds collected and plants grown in high density. From the nearly identical F_1 plants, seeds are collected. F_2 generation is grown from these seeds and one seed from each plant is collected. The F_3 generation is grown and one seed from each plant is collected randomly. This process is repeated for about two further generations. From say the F_6 generation, seeds collected from each plant of F_5 generation are sown with proper spacing, growing plants monitored and seeds harvested from plants with maximum heritable traits. These progeny seeds are grown in rows, undesirable off-type plants rogued off and seeds from selected progeny plants harvested and bulked. This process is repeated for few generations. After such intense effort, pure-line varieties are obtained.

Back cross breeding

This is a widely used method for introducing a gene of a good trait from one crop into another. There may be a crop X which is well adapted and high yielding but is susceptible to some biotic or abiotic stress like pest attack. Such a crop will be ideal if it becomes pest resistant also. There can also be a crop Y which is unadapted and low yielding but has the gene of pest resistance. By backcrossing the required pest resistant gene can be transferred from Y to X while filtering out the undesirable genes like those of low yield. When an adapted crop is crossbred with an unadapted crop, the progeny will have half of the total chromosomes from each parent, that is, a haploid set from the adapted crop and another haploid set from the unadapted crop. The purpose of backcrossing is to remove all unwanted genes of the unadapted crop but for the desired gene in the final progeny product. This is achieved by repeated *crossing* of the progeny of each generation *back* with the original adapted parent. Therefore, this protocol is termed as *backcrossing* (Fig. 8.13). The original adapted parent receiving the gene for improvement is referred to as the *recurrent* parent and the unadapted crop which is the source of the desired gene allele as the *donor* parent. After the first cross, in each generation large number of plants are grown to identify the progeny with the desired gene. Such progeny plants with the desired trait from the donor are selected and backcrossed with the recurrent parent. This process is repeated till the progeny is like the recurrent parent plus the desired trait. It takes several generations to achieve this, with the donor alleles reducing by ½ in each generation as determined by the relationship $1/(2^{N+1})^4$, where N is the number of backcrosses.

The products of above procedures are of sufficient homozygostiy to be considered as pure lines. Their seeds can be reused for three to four generations.

Cross Pollinated (Out Crossing) Crops

In cross pollinating crops, there is random mating in a stand of plants which results in ample genetic heterogeneity in the progeny. This is in obvious contrast to self pollinating crops. Although improvement of outcrossing species also involves repeating steps of selection, crossing and selection, the challenges are many because of unbridled natural crossing. Usually, to start with, base populations are developed for further breeding processes.

Mass and recurrent selections

For *mass selection*, plants of a crop species are grown, random cross pollination allowed and seeds of plants with suitable traits harvested.

This process is repeated for several generations till a satisfactory population is obtained. Seeds from such improved population of a crop can be processed for other breeding programmes like recurrent selection.

Recurrent selection is done by growing plants, intercrossing them and harvesting seeds of selected plants in the first generation. Seeds from each selected plant is bulked separately. Half of the seeds harvested are saved and the rest sown in rows for the second cycle, each row of seeds from the same plant. Random mating is allowed and again seeds from suitable plants from each row harvested. In the third cycle, seeds of plants selected in the second cycle and those stored from the first cycle are sown in rows. Plants of 2nd cycle are intercrossed with the recurrent parents of the 1st cycle, allowed to grow and again the desirable plants selected. The cycles may be repeated till a satisfactory basal population is obtained for further breeding programmes.

One use of such basal populations is the development of fodder grasses with prolonged vegetative phase. For such fodder grasses, inbred lines developed through mass or recurrent selection are crossed to develop *synthetic* varieties of fodder.

The crops developed by such methods are very useful in the third world countries as the seeds are reusable. These reusable and relatively less priced seeds are of great value to specially the small and marginal farmers who constitute majority of the farming community.

Hybrid varieties

The discovery of heterosis many years back led to breeding of hybrids. Heterosis or hybrid vigour is the phenomenon of an enhanced level of a trait in the progeny, better than either parent. A hybrid variety is produced, usually by crossing inbred lines of the chosen crop. The inbred lines are developed through carefully conducted mass or recurrent selection. These are stored for further use. For hybrid production the two parent inbred lines are sown in adjacent rows, one row is emasculated to act as female, allowed to mate and the seeds from the 'female' row harvested. As emasculation is a manual laborious process, male sterile inbred lines have been developed to facilitate hybrid production. For several years, such hybrid seeds are produced and subjected to rigorous testing before large scale commercial production.

Although hybrids have better performance, a great limitation is that the seeds cannot be reused. Resowing of F_1 seeds leads to segregation of traits in the F_2 generation.

The above accounts show that recombinant DNA technology or genetic engineering is necessary for production of GM crops. But plant tissue culture and plant breeding are essential enabling techniques.

8.8. TRANSGENES AND TRANSGENIC CROPS

Advances in recombinant DNA technology have led to a better understanding of the genome, the gene and identification of valuable genes. The following sections describe some useful transgenes and their applications. So far, the main genes engineered to produce successful GM crops have been of *input traits* as, resistances to insects, herbicides and pathogens. Other applications involve introduction of *output traits* as production of nutrients, vaccines or biofuels.

Insect Resistance

As, extensive use of pesticides constituted a major environmental hazard, attempts were being made to find alternative agents for pest control. This led to the initial discovery and use of certain genes. Thus, genes for protease inhibitor which interfere with insect protein digestion, chitinase which degrades the chitin of the cuticle cover of some insects and lectins which kill by binding with some crucial glycosylated insect proteins were found and used. However, the most effective gene was found to be the Bt endotoxin gene from the bacterium *Bacillus thuringiensis* (Bt). Sprays of this bacterium for pest control was being practised already. After the spray, when larvae feed on leaves, these bacteria are also ingested. In the midgut of the insect, the bacteria produce and secrete a toxin precursor, the protoxin. In the insect gut the protoxin is converted to the toxin and kills the pest.

The Bt toxin exists as protoxin in the *Bacillus thuringiensis* bacteria. The protoxin molecules form crystals during the sporulating phase of the bacterium and are therefore termed as *cry* proteins. In the insect larvae guts, the alkaline medium degrades the protoxin molecule to the toxin. These *cry* molecules bind to specific receptors in the epithelial membrane of the insect gut and oligomerize to form many pores in the wall. This makes the gut porous and leaky leading to starvation and death of the insect larvae. The cry protein-receptor binding is very specific. Therefore, a *cry* toxin which may be fatal to one group of insects may not affect insects of other groups. For instance, the *cry* proteins of the *cry1A* gene affect only members of Lepidepteran insects (moths and butterflies), those of *cry3A* mainly Coleopterans (beetles). Similarly, there are specific *cry* proteins for Diptera (flies and mosquitoes). Bt corn and Bt cotton are the most widely used insect resistant GM crops. To avoid early development of resistance to Bt toxin, attempts are being made to produce GM cotton and corn with stacked multiple *cry* genes fatal to the attacking pest.

Herbicide Resistance

In countries like the US, land holdings of the farmers are very large as compared to the third world countries. Manual weeding of huge tracts of farm land becomes highly labour intensive and financially untenable. Therefore, weeding is done by herbicide sprays. It is necessary therefore, to have crop plants which are resistant to the herbicide being sprayed. There are two types of such herbicide resistant plants the *glyphosate resistant* and *glufosinate resistant* plants.

Glyphosate interferes with the synthesis of aromatic aminoacids as tyrosine, phenylalanine and tryptophan by the shikimate pathway. The herbicide inhibits the enzyme EPSPS (5-enolpyruvylshikimate-3-phosphate synthase) of the shikimate pathway. Due to this interference, the aromatic amino acids are not formed and the plants die. Scientists discovered in some bacteria a glyphosate resistant form of EPSPS. Glyphosate resistant soybean (Roundup Ready, RR soya) and corn were developed by transferring this gene to these crops. As the shikimate pathway works in the chloroplast, a leader sequence is added to the gene to ensure that the enzyme coded has an attached transit peptide sequence, to enable it to be transported to the chloroplast. These crops are being widely used in the US.

The other herbicide, glufosinate acts by inhibiting the enzyme glutamine synthase, which is responsible for glutamine synthesis. Inhibition of this enzyme by the herbicide, checks glutamine synthesis leading to ammonia accumulation in the plant with subsequent death. Luckily, it was found that some *Streptomyces* bacteria produce glufosinate probably as an evolutionary advantage, as this compound has both phytotoxic and antibacterial properties. At the same time, some bacteria resistant to glufosinate were also discovered. Studies revealed that resistance of these bacteria is because of a gene which produces the enzyme, phosphoinothricine acetyl transferase (PAT). This enzyme encoded by a gene, the *bar* gene, degrades glufosinate. The bar gene isolated from *Streptomyces hygroscopicus* has been transferred to several crops to generate glufosinate resistant GM crops (LibertyLink series).

Disease Resistance

Crops are always subject to attack by microbes as fungi, bacteria and viruses. Through conventional breeding many pathogen resistant plants have been developed. However, there are difficulties like the resistance gene may often be linked to an undesirable gene and thus, it may be impossible to transfer it independently. The advent of recombinant DNA technology has opened up some possibilities. Natural resistance of a plant to a pathogen is because of a resistance gene, the R gene. This gene is activated on pathogen attack. When a pathogen invades a plant it produces some proteins coded by its *Avr* (avirulence) gene. The R gene also gets activated and produces proteins which recognize and interact with the *Avr* proteins and induce the defence process. This phenomenon could be simulated to produce pathogen resistant GM crops. A classic example is the development of bacterial blight resistant GM rice. Normal rice is highly susceptible to *Xanthomonas oryzae*. GM rice resistant to this pathogen has been developed by transferring the R gene *Xa21* from a wild resistant rice variety *Oryza longistaminata*.

One of the mechanisms of tackling viral infections in many organisms as plants and even humans is *cross protection*. Often, after a cycle of infection by a virus, the host develops resistance to it. The host does this by producing RNA complementary to the invading RNA and thus silencing it by hybridization. An interesting mechanism governs this process of resistance. There is an enzymatic RNA induced silencing complex (RISC) in these organisms. When a virus attacks and transcribes coat protein RNA, the RISC system gets activated, recognizes the viral RNA, and silences it by producing complementary RNA. This process has also been simulated to produce some GM crops resistant to particular viruses. For instance, part of a coat protein gene of a mild strain of PRSV (Papaya ring spot virus) was introduced in papaya. The transcription of this coat protein RNA, activated the inherent RISC system leading to further production of RNA similar to the RNA of invading PRSV. The host RNA silenced the coat protein RNA of invading PRSV. The pathogen died as the coat proteins could not be synthesized.

High Nutrition Value

Attempts are also being made to produce transgenic plants enriched with some valuable nutrients as vitamins or proteins. Thus, increasing the protein content of otherwise carbohydrate rich potato makes it a more wholesome food. An example of enhancement of the nutritive value of a widely consumed crop is Golden rice. This GM crop also illustrates possibilities of multigene transfer. Rice is a staple food for most communities in the South Asian countries. Normal rice grains have an aleurone layer surrounding the endosperm. This layer is rich in vitamins, including vitamin A. The aleurone layer is retained in *parboiled* rice, but, in the normally consumed *polished* rice this layer gets stripped off. Naturally, polished rice is of lower nutritive value. Through genetic engineering, however, a GM rice, the Golden rice has been produced which has its endosperm enriched with β-carotene, a precursor of vitamin A.

Carotenes are terpenoids, built of chains of five carbon isoprene units. Rice tissues including developing grains, have the main 20 C precursor of β-carotene, geranylgeranyldiphosphate (GGPP). Betacarotene synthesis takes place by union of two GGPP units by *phytoene synthase* to form phytoene, its desaturation to lycopene by desaturase and conversion of lycopene to β-carotene by *lycopene cyclase*. Thus for making Golden rice, genes for these three enzymes were introduced and 'stacked' in rice through the process of *cotransformation* by particle bombardment. Initially phytoene synthase and the cyclase were taken from daffodil and desaturase from a bacterium of *Erwinia* sp. As the carotene content was low in the first formed Golden rice, further efforts were made. Soon *Golden rice 2* was developed with phytoene synthase from corn rather than daffodil. The β-carotene content in this later version was about 20 times higher than the original Golden rice.

There are several other applications also involving output traits. Some prominent ones are transgenic plants for pharmaceuticals as oral vaccine, biofuels, modified oils and plants of higher nutritive value like Golden rice mentioned above.

8.9. TESTING OF GM CROPS: TRANSGENE ANALYSIS AND FIELD PERFORMANCE

The GM crops are tested rigorously at two levels. After production of the primary transformant or event, the first tests are to check for proper transgene insertion and functioning, by molecular techniques. The next crucial tests are after transfer of the transgene to an agriculturally suitable variety for field performance. Besides the performance tests of conventional plant breeding, the GM crops are additionaly tested for biosafety as per the guidelines of *Codex Alimentarius*. The biosafety checks include studies on environmental impacts and in case of food crops, further tests as *substantial equivalence* with the untransformed parent variety.

Transgene Analysis

During the transformation process several integrations take place each being an *event* of transformation. Transgene analysis is done to select the transformation event exhibiting a single insertion and high gene expression. Such a single gene high expression integration and the resulting transformant progeny is termed as a *transformation event*.

The insertion of the transgene and its functioning is tested by established tools of molecular biology. Transgene DNA integration is checked mainly by Southern Blot hybridizaton with probes and PCR. Transgene functioning is tested at two levels, *transcription* and *translation*. Transcription level of the transgene is assessed by Northern Blot and also reverse transcriptase PCR. As, expression of the desired trait of the transgene is dependent on the adequate synthesis of the protein it codes, it is necessary to evaluate the transgene at the translation level too. The tests are mainly based on the Western bloting technique. Before the detailed tests, a screening is done to select the probable or *putative* transgene containing transformants.

Screening of putative transformants

The transformants obtained initially constitute a population of probables with the promise of there being at least one ideal single gene high expression event. Such transformants are referred to as *putative* transformants. This population will contain a mix of different kinds of integrations varying in the copy number and positions of the transgene integrated. All such explants are grown independently to maturity and self pollinated to produce seeds. Seeds from each transformant is collected on agar medium containing the selection agent *e.g.*, an antibiotic like kanamycin or hygromycin, against which the resistance gene exists in the transformant. The transformant containing a single gene in hemizygous state on a chromosome will segregate in a ratio of 3:1 :: survivor:nonsurvivor. Such *single copy* transformants are selected for futher molecular studies. These tests have to be reinforced by tests for the presence of the transgene by simple PCR. A common simple test is for the protein coded by the transgene. This is done by the ELISA (enzyme linked immuno sorbent assay). A bulk protein extract is made and is pipetted into wells containing a previously prepared antibody against the transgene coded protein. The antibody-transgene protein complexing occurs and is detected by addition of an enzyme which also conjugates with the protein bound antibody. This complex is detected by washing with the appropriate substrate.

The transformant selected on the basis of these studies are subjected to more definitive tests as described below.

Determination of copy number of transgene insert

One useful method is Southern blot analysis (chapter 4). The total DNA is extracted and fragmented with a restriction enzyme, the recognition site of which does not exist in the inserted DNA. The digest is fractionated by suitable gel electrophoresis, the agar plate is processed for Southern blotting. After transfer of the bands to the membrane they are examined with a labeled probe complementary to the transgene DNA. The probe can be a PCR copy of the transgene or cDNA made from the transgene's mRNA transcript. Occurrence of a single band indicates one copy single gene insertion. Whether the entire gene with promoter-coding region-terminator has integrated or not is revealed through more specific blotting experiments.

Determination of the transcript mRNA

The presence of the complete transgene DNA is no guarantee that it is being expressed. Actual presence of the mRNA transcript of the transgene, in good amounts, in the transformed tissues confirms proper functioning of the insert. The mRNA transcript is detected by Northern blotting. Total mRNA is extracted and fractionated by gel electrophoresis and the gel is blotted on a nylon membrane. The transcript is identified by labeled single stranded complementary DNA or RNA.

Determination of the translation polypeptide product

The main method of detection is Western blotting. Total protein is extracted from the transformed tissue and fractionated by polyacrylamide-SDS gel electrophoresis. The protein bands are transferred to a polyvinylidine fluoride (PVDF) membrane. The bands are detected by a sensitive antibody probe assay. For this, prior to the start of blotting, a stock of antibody, specific to the transgene protein product is prepared. The gel with protein bands is bathed in a solution of the antibody. The antibody binds only to the transgene coded protein.

The gel is washed to remove any unbound antibody. The bound antibody is detected by treating the gel with an enzyme conjugate secondary antibody which conjugates with the specific primary antibody resulting in chemiluminiscence or colour change.

For simpler rapid tests, the translation product can also be detected by subjecting a protein extract of the tissue to ELISA test.

Phenotype Analysis

Once it is confirmed that the transgenic product has the proper dose and quality of the transgene, it is necessary to know the impact it may have on other phenotypic traits of the plant. The insertion of transgene should not alter any other phenotypic trait. Therefore, a comparative study of phenotypic traits of the transgenic and the original parent line should be conducted. The parameters should include, seed germination, seedling emergence, growth performance, flowering, seed set, yield, disease resistance and any other trait deemed necessary. An ideal transgenic plant should be similar to the original parent plant but for the trait introduced by the transgene.

Field Tests

Besides the laboratory tests the GM crops are subjected to rigorous field tests also before release. There are two sets of tests, a) agronomic tests of growth, yield, resistance to abiotic and biotic stresses as compared to the original parent or isogenic line, b) biosafety assessment involving environmental risk assessment and food and feed safety in case of food crops. The assessments also depend on the extent of environmental exposure and the regulatory controls at the international and national levels.

One simple test for checking transgene flow again is, ELISA on *e.g.*, leaves or seeds of non GM crops growing in the proximity of GM crops. ELISA is a tool for testing transgene / transgene product presence in a material as seeds of non GM crops growing near a GM crop field. It is also widely used in areas as medicine and plant pathology and for quality control checks in various industries. For this test, antibodies are raised against an antigen as a synthetic transgene. The sample extract with suspected transgene contamination is attached to a matrix surface. Then, the sample matrix is bathed with the specific antibody so that it can bind to the antigen if present. For detection assay, the antibody is linked to an enzyme. If the transgene or its product is present the antibody binds to it. The unbound antibody molecules in the assay mixture solution are washed away and the enzyme linked bound antibody is detected by adding the enzyme's substrate. The presence of the contaminating transgene is detected by a signal, most commonly a color change in the substrate. Absence of signal indicates absence of transgene contamination.

Agronomic assessment is based on time tested protocols of performance evaluation with wide range of filed trials. For environmental and food safety, the usual steps are:

(i) assessment of the needs, that is items which may constitute potential hazards and need to be tested,

(ii) designing of laboratory experiments, analyses and field trials for generating meaningful data,

(iii) performing the experiments and trials as per the designs formulated, and

(iv) statistical analyses and interpretation of the data generated.

9

Transplastomic Plant Production by Chloroplast Transformation

SUMMARY OF CONTENTS

The plastid genome; chloroplast transformation; applications of chloroplast transformation in functional genomics; transplastomic plants of economic importance

The transgenic plants in use at present have been produced by nuclear transformation. This method which is being used widely for commercial production of GM crops has several inherent problems. The transgene insertion is random and leads to position effects with consequent varying levels of expression. There are always possibilities of gene transfer from the transformed crops to untransformed crops in adjoining fields by pollination and similar gene contamination of wild relatives. Further, usually, only one gene can be introduced at a time. The plants produced by chloroplast transformation are termed as *Transplastomics* and the transgene is located in the chloroplast genome rather than the nuclear genome. Chloroplastic transformation has several advantages. In this method the gene is inserted by targeting to a predetermined location and the integration is not random. In a cell there is only one nucleus containing one diploid genome. In contrast, a cell has many chloroplasts, each containing several genomes. Thus, while a nuclear transformed plant usually contains only one transgene per cell, transplastomics can contain many transgenes per cell. This leads to synthesis of higher amounts of the transgene encoded proteins in transplastomics than in the nuclear transformed plants. A major advantage is the maternal inheritance of the chloroplast borne transgene. The transgene being present in the chloroplast is passed on to the progeny through the egg cell during fertilization (involving fusion of the female nucleus of the egg cell and the male nucleus from the pollen).This is because while the egg cell, besides the haploid nucleus, contains all the cytoplasmic components including the chloroplasts, the pollen has only the nucleus. So, the organelles including chloroplasts are maternally inherited and so are the genes borne by them. As the pollens do not have the transgene containing chloroplasts, there is no chance of gene contamination by horizontal gene transfer (HGT) through pollination (Fig. 9.1).

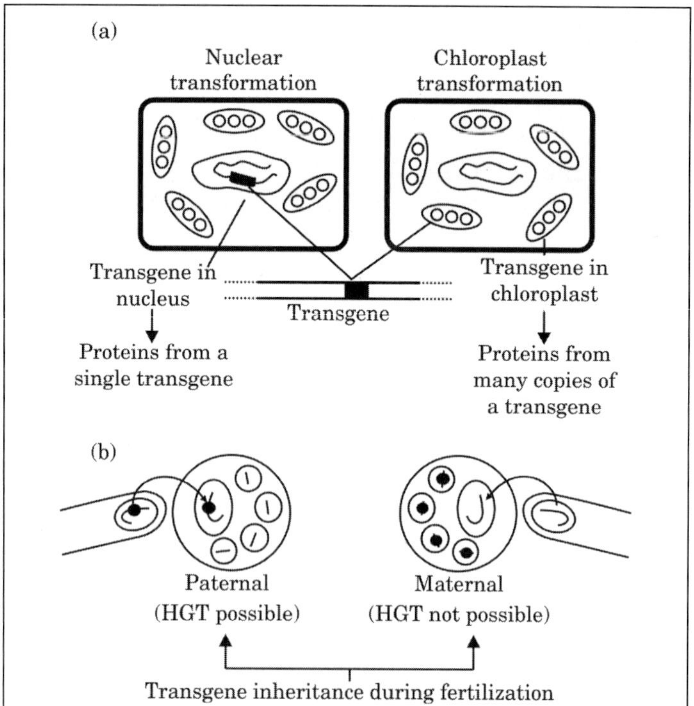

Fig. 9.1: Nuclear and chloroplast transformations, salient features, (a) More protein product in transplastomics, (b) No possibility of gene contamination through horizontal gene transfer (HGT)

The chloroplasts are self contained in post transcriptional processing of the proteins formed, like, assembly of subunits, -S-S- bonding and folding of the polypeptide chains formed. The foreign proteins encoded in the chloroplasts are also much more stable than similar nuclear encoded proteins. Since the chloroplasts have a polycistronic gene expression mechanism, an additional advantage with transplastomic technology is the possibility of multigene transfer in a single transformation step.

9.1. THE PLASTID GENOME

There are different types of plastids. The main categories are the chlorophyll containing *chloroplasts,* the *chromoplasts* with other pigments as lycopene, the starch containing *amyloplasts* and the oil containing *elaioplasts.* Proplastids are developing plastids. The plastid genome or the plastome is highly conserved with common features in all plastid types. The genome in each plastid consists of copies of circular DNA nucleoids. In a mature cell, there may be 100 chloroplasts, each chloroplast with around 100 nucleoids. Thus a cell can have 10,000 copies of the nucleoids. This is the reason for high level production of transgene encoded protein in chloroplasts, sometimes about 45% of the total cell soluble protein.

The plastome nucleoid is a circular DNA, 120 to 160kb long, with two inverted repeats, and containing about 130 genes (Fig. 9.2). The genes are organized in operons and gene expression is mainly polycistronic. In this way, the plastome is like its prokaryotic ancestors. An example is the barley plastome, 136462 bp long, with a large single copy (LSC) segment 80695 bp long and a small single copy segment (SSC) of 11724 bp separated by 22021 bp inverted repeats (IR).

It is believed that the chloroplasts originated by the initially evolved cells engulfing some prokaryotic cyanobacteria. The endosymbiont cyanobacteria gradually merged with the eukaryotic cellular system. This evolution involved the transfer of about 3000 genes from the endosymbiont to the cellular nucleus over time.

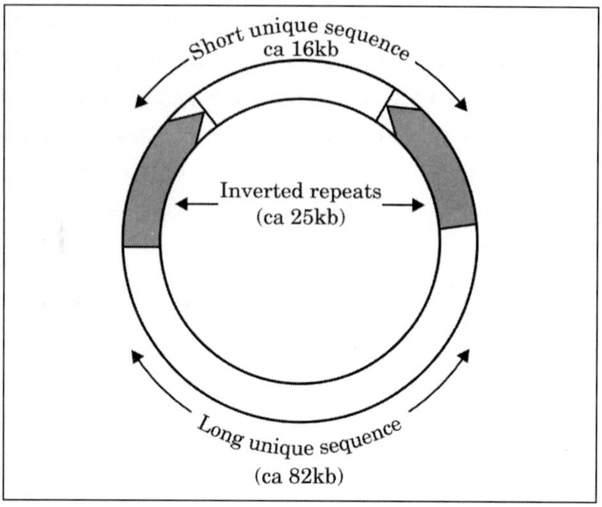

Fig. 9.2: A generalized scheme of the chloroplast genome

9.2. CHLOROPLAST TRANSFORMATION

Initial success in chloroplast transformation was with the green alga *Chlamydomonas reinhardtii,* followed by higher plants as, tobacco, potato and tomato. To the genetic engineers, the double membrane of the chloroplasts and difficulties in designing of suitable vectors in the absence of parasitic microbes for chloroplasts, created some barriers for gene insertion into the plastome. However, invention of the biolistic protocol helped. Now the most common method for chloroplast transformation is the gene gun or biolistic method.

In general, two distinct components are required to construct the final chloroplast transformation vector system, a vector containing the flanking sequences, left flanking sequence, and right flanking sequence, and the sequences required for efficient transgene expression.

As has been mentioned above, the transgene integration is by homologous recombination. The transgene is inserted in a space between genes, an intergenic region. Therefore, the main components of the vector are kept between two plastome intergenic sequences, the flanking sequences. Since the sequences of the plastomes are mostly known, it is easy to choose such sequences. A vector for chloroplast transformation consists of, markers, promoters, ori regions, a cloning site into which the gene of interest is inserted and other elements, flanked by two chosen intergenic plastome sequences (Fig. 9.3).

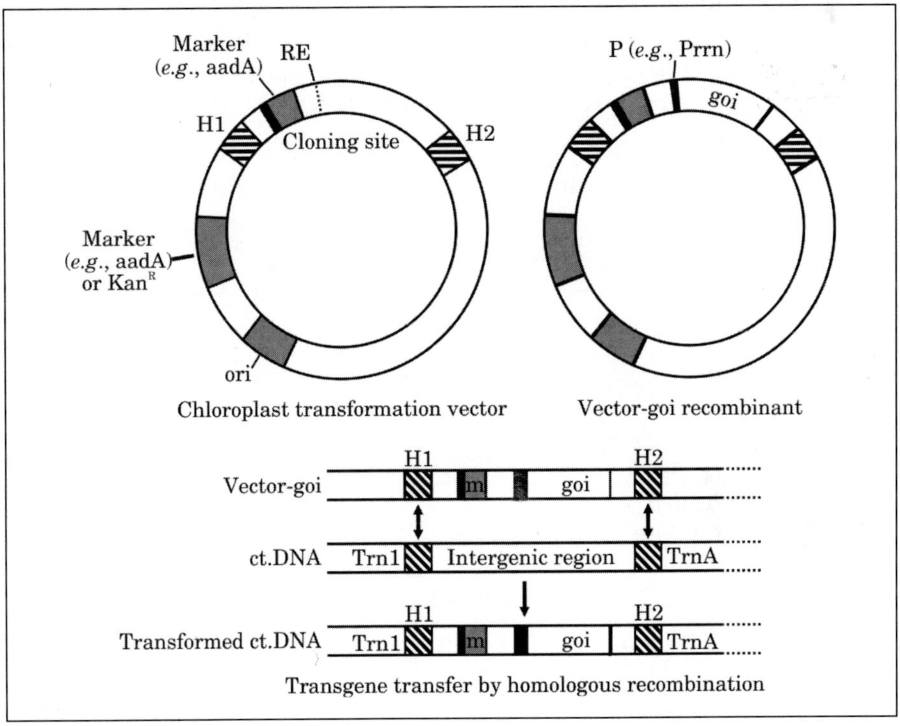

Fig. 9.3: Chloroplast transformation vector, recombinant and transgene transfer (H1 and H2 = sequence homologous to, chloroplast DNA intergenic region between *e.g.,* trn1 and trnA genes)

The ori sequences are mainly for replication in *E. coli* (for cloning). The commonly used marker genes are for spectinomycin, streptomycin and kanamycin resistance. For resistance against streptomycin and spectinomycin, the *aad*A gene, which encodes the detoxifying enzyme aminoglycoside-3′-adenylyl transferase (aad) is used. For kanamycin resistance, a gene of the *aph* family encoding aminoglycoside phosphotransferase is used. The preferred promoters are of chloroplast origin *e.g.*, the Prrn promoter. Endogenous regulatory elements and flanking sequences are essential for efficient chloroplast expression. In the biolistic method, the transgene-vector recombinant DNA is coated on gold or tungsten particles. A common vector is the plasmid derived pZS148 vector. These particles are shot into leaf discs, calli or cell cultures in a helium pressure aided gene gun and selected thereafter for transformed cells/tissues. Initially, the transgene integrates with very few nucleoids of the plastome. Therefore, there is both intra and interplastidic heterogeneity or heteroplasmy. There is a mix of transformed and wild type nucleoids in a plastid and transformed and wild type plastids in a cell (Fig. 9.4).

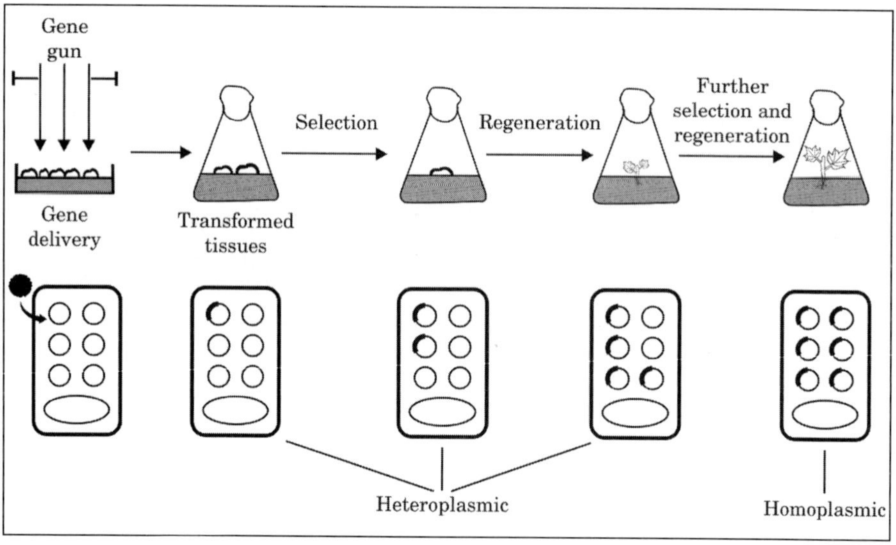

Fig. 9.4: Steps of chloroplast transformation and formation of homoplasmic transplastomics

By repeated selection and regeneration of transformed cells or transformed excised tissues, totally transformed homoplasmic explants are obtained (Fig. 4). Homoplasmy is the stage when all the nucleoids of each plastid and all the plastids of a cell contain the transgene or DNA element introduced. In Table 9.1 are shown few particulars of some transgenes introduced in transplastomic plants.

Table 9.1: Transgenes in transplastomics

Trait	***Gene***	***Integration site***	***Promoter***	***5'/ 3' UTRs***
		Agronomic Traits		
Pest resistance	*Cry*2Aa2	*rbc*L/*acc*D	Prm	ggagg/T*psb*A
Hebicide resistance	*bar*	*rbc*L/*acc*D	Prm	*rbc*L/T*psb*A
Disease resistance	*MSI*-99	*trn*I/*trn*A	Prm	ggagg/T*psb*A
Drought tolerance	*tps*	*trn*I/*trn*A	Prm	ggagg/T*psb*A
Salt tolerance	*badh*	*trn*I/*trn*A	Prm-F	ggagg/T*rps*16
		Genes Encoding Medicinal Products		
Human somatotropin	*hsT*	*trn*V/*rps*12/7	Prm	T7gene10/T*rps*16
Interferon gamma	IFN-g	*rbc*L/*acc*D	P*psb*A	P*psb*A/T*psb*A
Monoclonal antibodies	*Guy's*13	*trn*I/*trn*A	Prm	ggagg/T*psb*A
Tetanus toxin	*Tet C*	T*rnv*/*rps* 12/7	Prm	T7gene10/T*rbc*L

To confirm integration of the transgene in the plastome, various methods are used. A reliable method is detection by PCR of the transgene in a pulse-field gel electrophoresis plastid DNA sample.

As mentioned above, there are many copies of the transgene in a cell. Therefore, the production level of the protein encoded by the transgene is much more in a transplastomic than in a nuclear transformed GM plant. Further, due to maternal inheritance the transgene is contained. It is not spread to other plants through pollination.

Marker free transplastomics can also be obtained. One of the successful methods uses the Cre/lox system (Fig. 9.5). The marker gene is placed in the vector with lox sequences on either side. The Cre gene is inserted in the nuclear genome. After transformation, the Cre gene is induced to synthesize a plastid targeted recombinase which excises the lox sites and thereby removes the marker. Another site-specific phage recombination system is the phiC31/*attB*/*attP* combination (Lutz *et al.,* 2007).

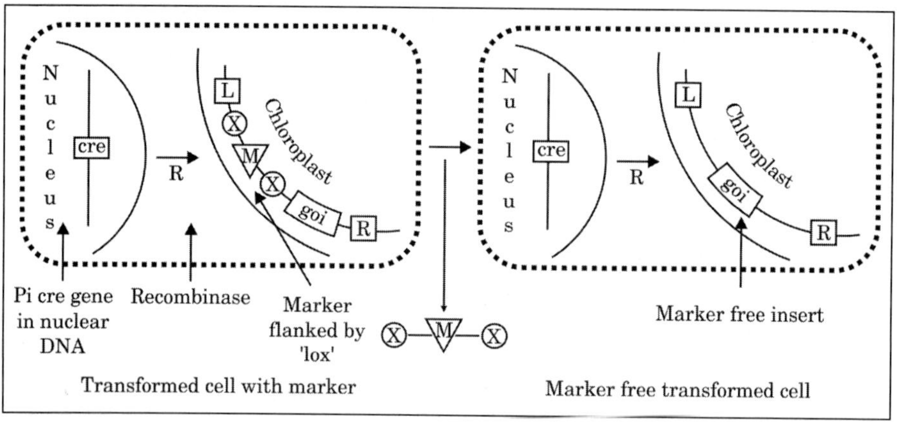

Fig. 9.5: Removal of marker by Cre/lox system for marker free transplastomic production (hypothetical scheme, Cre = P1 recombinase gene, X = lox site, L and R =left and right repeats, goi = gene of interest)

An emerging selection marker is the betaine aldehyde dehydrogenase (*badh*) gene. This gene occurs in plants as spinach and those adapted to drought or salinity. The enzyme converts poisonous betaine aldehyde to nonpoisonous glycine betaine. As this gene does not occur in all plants, it can be used as a selection marker. The transformed plants are selected by exposure to betaine aldehyde rather than antibiotics. The added advantage is that glycine betaine is an osmoprotectant and therefore, the *badh* gene can be used to make drought resistant and salinity resistant transplastomic plants.

Chlamydomonas reinhardtii and tobacco were the first transplastomics produced. Thereafter, chloroplast transformation has been achieved in several other plants as potato, tomato, carrot, cotton and soybean. Chloroplast transformation in monocots has been difficult.

An alternative approach of plastid transformation technology for other recalcitrant species of *Solanaceae*, has been the use of remote cytoplasmic hybrids. This method was based on two somatic hybridization steps (Kuchuk *et al.,* 2006). Initially, two remote hybrids (cybrids) carrying the nucleus of tobacco, an easily transformable species, and plastids of the recalcitrant *Solanaceae* species were used. Cybrid protoplasts were then subjected to PEG-mediated plastid transformation. Later, protoplasts (with defective nucleus) from cybrid transplastomic plants were asymmetrically fused with protoplasts of the recalcitrant species, which originally provided the plastids in the first somatic hybridization step. By this technique, successful genetic transformation of plastids of five species of *Solanaceae* could be achieved.

9.3. APPLICATIONS OF CHLOROPLAST TRANSFORMATION IN FUNCTIONAL GENOMICS

Chloroplast transformation emerged as a major tool for identifying plastid genes involved in photosynthesis and other genes, besides various gene regulatory elements.

Reverse Genetics by Gene Knockouts

The normal approach in genetics has been to start with a phenotypic trait and through appropriate studies identify the gene concerned. This traditional approach has been termed as forward genetics. Sequencing studies have inspired a newer approach of starting with a DNA sequence, a probable gene, and determining the trait it controls. This reverse genetics is conducted by studying the phenotypic effects of gene knockouts, gene deletions or modifications.

One approach is replacement of the target gene, knocking out the sequence by a selection marker gene through homologous recombination and analyzing the effects. Another method is gene deletion by the Cre/lox system. An interesting example is the *clpP1* gene. *In vitro* studies indicated that this gene encoded the ClpP1 protease subunit. This finding raised the query about the role of this protease and therefore the gene. Deletion of the *clpP1* gene by Cre/lox mechanism led to stem growth indicating a role of this gene in stem development via the encoded protease.

Expression Enhancement

It may be necessary to enhance the expression of a gene to obtain requisite amounts of a product. Several examples of such chloroplast transformation mediated transgene overexpression exist. The *accD* gene could be overexpressed by replacing its native promoter with the rrn promoter. Similarly, shifting the anthranilate synthase gene from the nucleus to plastome, enhanced its expression with consequent higher levels of tryptophan synthesis.

Transcription Analysis

Plastid transformation has helped in revealing precise functions of genes and the associated regulatory elements. The role of RNA polymerases encoded by plastids and the nucleus have been understood by knocking out the genes encoding these enzymes. Removal of the *rpoA* and *rpo*B genes for the plastid encoded polymerase (PEP) and the genes for the nuclear encoded polymerase (NEP) followed by analysis of the respective knockout plants showed that these enzymes transcribe different groups of genes. In other studies, splicing of PEP and NEP associated sequences with reporter genes as *gus* and *gfp* led to the discovery of promoter elements of these enzymes.

RNA Editing

In plastids, there is a unique post transcriptional RNA editing system. In certain regions of the transcribed mRNA, conversion of cytosine to thymine (C→T) occurs. Only after such transition the mRNA can be translated. To identify the editing elements, short mRNA fragments transcribed from such editing DNA sections were fused with reporter gene mRNA and processed for translation. Successful translation indicated the occurrence of correct editing elements. Such studies indicated the involvement of a 22 nucleotide sequence spanning the editing site.

Genes Involved In Photosynthesis

Gene knockouts have revealed the role of many genes of the photosynthetic machinery. Further, attempts for improving photosynthetic efficiency have also been made. One such approach has been the manipulation of the genes encoding Rubisco enzyme which is responsible for CO_2 fixation. Rubisco has a small and a large unit. The gene *rbc*L encoding the large unit is in the plastid and the gene *rbc*S for the small unit is in the nucleus. Replacement of the *rbc*L gene of tobacco by heterologous genes from sunflower and some other organisms seemed to enhance photosynthetic efficiency. The role of several open reading frames (ORF) were also determined including different ycf sequences (h*y*pothetical *c*hloroplast reading *f*rame). Each ycf was isolated, mutated and used to replace the wild type ycf by homologous recombination. Subsequent phenotypic analysis revealed the role of a particular ycf. Such studies showed for example, the need of ycf3 and ycf4 for PS1 assembly and of ycf6 for the proper organization of the cytb6f complex.

9.4. TRANSPLASTOMIC PLANTS OF ECONOMIC IMPORTANCE

Many transplastomic plants with useful traits have been produced. Because of the ability of the transplastomics to produce high levels of encoded products, they are being used also as biofactories for production of compounds of medicinal value and other products. A brief account of some of the transplastomics with economically important traits is given below.

Insect Resistance

Tobacco plants have been transformed with the Bt toxin *Cry2Aa2* operon from a sub species of *Bacillus thuringiensis.* The bacterial gene works better in the chloroplastic ambience. After synthesis, the chaperone activity of the accompanying *orf 2* gene folds the *Cry2Aa2* toxin protein into cuboidal crystals. The leaves of such plants were highly toxic to the attacking pests as tobacco bud worm, *Heliothis virescence* (De Costa *et al.,* 2002).

Herbicide Resistance

A common herbicide is glyphosate which acts by inhibiting a key enzyme EPSPS (5-enolpyruvateshikimate-3-phosphate synthase) of the aromatic acid biosynthesis pathway. The resistance is conferred by inserting a heterologous EPSPS wich is insensitive to glyphosate and overexpresses. Nuclear transformed glyphosate resistant plants are effective, but, there is always the chance of spreading herbicide resistance to weeds via HGT through pollen. Therefore, sequestering the transgene in the chloroplast and keeping it under the control of maternal inheritance is safer. Transplastomic glyphosate resistant plants were produced by inserting a wild type petunia *aroA* gene in the *trn*I/*trn*A intergenic region of the invert repeat or *rbc*L/*aac*D intergenic space of the large single copy segment of the plastome.The plants showed 10 fold greater resistance (Daniell *et al.,* 1998).

Wurbs *et al.* (2007) produced transplastomic tomato expressing bacterial lycopene β-cyclase gene, resulting in increased levels of herbicide tolerance to 2-(4-chlorophenylthio)-triethylamine (CPTA), which specifically inhibits lycopene β-cyclase activity.

Pathogen Resistance

A broad spectrum antimicrobial agent is MSI-99, a helical peptide. This peptide has high affinity for negatively charged phospholipids which are abundant in the microbe membranes. On contact with a microbial cell it induces aggregation of the membrane phospholipids leading to pore formation, disruption of the membrane and death of the cells. Transplastomic tobacco transformed with MSI-99 gene were resistant to *Pseudomonas syringae* cv *tabaci* and *Colletotrichum* (De Gray *et al.,* 2001).

Drought, Salt and Cold Tolerance

Drought resistant plants have the inherent potential of synthesizing osmoprotectants in response to abiotic stresses for their protection. These are compounds like sugars, alcohols, proline and some nitrogenous compounds. Trehalose is such an osmoprotectant and its synthesis is mediated by the enzyme trehalose phosphate synthase. This enzyme is encoded by the enzyme *TPS1*. Attempts have been made to transform these plants with the *TPS1* gene to confer drought resistance. Nuclear transformation led to low expression levels and pleiotropic effects. However, chloroplast transformation with this enzyme led to successful drought resistant plants with high levels of trehalose in plastids and without any pleiotropic effects (Lee *et al.,* 2003).

In some plants, glycine betaine accumulates as an osmoprotectant against salinity stress. The synthesis of this compound is mediated by the badh enzyme. Kumar *et al.* (2004) developed salinity resistant carrot transplastomics by engineering the *badh* gene in the chloroplasts of these plants.

Craig *et al.* (2008) produced transplastomic tobacco plants, which express a *Delta-9 desaturase* gene from wild potato species *Solanum commersonii*. This gene controls the insertion of double bonds in fatty acid chains, and the resulting transplastomics had altered leaf fatty acid profiles and increased cold tolerance.

Phytoremediation of Mercury Ions

Ruiz *et al.* (2003) developed transplastomic tobacco which could survive in high concentrations of phenyl mercuric acetate (PMA) and also accumulate mercury. This was achieved by engineering the two bacterial enzymes, mercuric ion reductase (*mer*A) and organomercurial lyase (*mer*B) in an operon in tobacco chloroplasts.

Besides the above traits, attempts are on to confer other traits also as cytoplasmic male sterility in a wide range of crops.

Transplastomics in Human Health and Medicine

Because of higher levels of production and built-in mechanisms of post translational processing of the proteins encoded, chloroplasts have very good potential as bioreactors for pharmaceuticals, vaccines and other materials as useful enzymes.

Human serum albumin (HSA) is in much demand for various therapies. The human resource is limited and can be a source of infections also. Production in transplastomics has been tried with some success. Chloroplast is specially suitable as HSA is AT rich and has many disulphide bonds. In the chloroplast, higher AT content does not limit protein synthesis and the disulphide bonds are formed and the protein folded as a post translational processing step. An HSA producing tobacco transplastomic has been made by inserting the *HSA* gene between the *trnA*/*trnI* space in the *rrn* operon and inducing its expression under the control of *psbA* regulators. The HSA production in the leaves is about 10% of total soluble protein (Fernandez-SanMillian *et al.,* 2003).

Strategies for the production of Human Interferons α and γ, some monoclonal antibodies as Guy's 13 antibody for the surface antigen of the dental caries causing *Streptococcus mutans* and other products are being developed as well.

Vaccination is practised world over for prevention from pathogen attack. In India, fatal pathogens as small pox virus have been eliminated due to effective vaccination programmes. The other common instance is the promotion of polio oral vaccine. The process is a facilitation of the natural process of immunity development.

When a pathogen enters the blood circulatory system, a chemical configuration on its outer membrane surface is recognized by the inherent immune system and antibodies are produced and released in the blood. These antibodies kill the pathogen. Continuing presence of the antibody imparts immunity against further attacks by the pathogen (This system works for pathogens like some viruses, bacteria and protozoans as long as the there is no variation in the pathogen). The infective agent is termed as the antigen and the basis of immunity is antigen-antibody interaction. Artificial immunity is induced by injecting the antigen, that is the pathogen, in an avirulent inactive form. So, a vaccine is a preparation of the pathogen in avirulent and inactive form as antigen. To avoid any risks, even remote ones as chance reversal of avirulence, it is better to use the main surface chemical in a pathogen recognized by the immune system, as the antigen. Recombinant DNA technology, specially chloroplast transformation, can be useful in obtaining the vaccines in good amounts.

Many endeavours have led to some successes. Vaccine against cholera has been manufactured in tobacco chloroplasts by engineering in an antigen portion of *Vibrio cholera,* the *CTB* gene without the leader sequence. The gene encodes cholera toxin beta subunit, a powerful adjuvant. Use of enhancer elements led to the production of the antigen to about 30% tsp levels (Molina *et al.,* 2004). The bacterium *Yersinia pestis* causes bubonic, pneumonic and septicemic plagues. An effective vaccine in the form of F1/V fusion protein from the bacterial membrane has been synthesized in tobacco chloroplasts upto 10% tsp level. There are several other examples of production of valuable vaccines.

Waheed *et al.* (2011) developed a cost-effective protocol for VLP-based HPV vaccines. A modified *HPV-16L1* gene, which retained the ability to assemble L1 protein to capsomeres was expressed in tobacco chloroplasts. Capsomeres are relatively thermostable and can induce the immunogenicity to a level as that of VLPs (virus-like particles).

The chloroplast transformation system has also been explored for the production of HIV antigens. Transplastomic tobacco plants were able to accumulate human immunodeficiency virus type 1 (HIV-1) p24 protein up to about 2.5–4.5% of tsp with correct size and without any post-translational modifications such as glycosylation or phosphorylation (McCabe *et al.*, 2008).

Transplastomics and Biomaterials

Genetic engineering of chlorplasts for production of important biomaterials at industrial level is also being tried. Some headway has been made in certain cases as liquid crystal polymers, biodegradable plastics, anthranilate synthase and some other materials. An important example is of xylanase production. Xylanase is extensively used in many industries as, paper, fibre, baking and others. The *xynA* gene integrated in tobacco chloroplasts expressed xylanase to the extent of 6% tsp, a viable level for commercial venture (Leelavathi *et al.,* 2003).

Bohmert-Tatarev *et al.* (2011) reported PHB (poly hydroxy butyrate) expression of up to 18.8% dry weight of leaf tissue, with use of an optimized gene construct based on their similarity to the codon usage and GC content of the tobacco plastome. PHB is a raw material for biodegradable plastics.

Chloroplast genetic engineering has a wide range of applications and gives us a reason to be optimistic as the technology is biosafe. However, certain limitations have to be overcome. The major hurdles to be overcome are low plastid transformation efficiencies of some of the species and inefficient gene expression in non-green plastids.

10

Intellectual Property Rights and Biosafety Regulations

SUMMARY OF CONTENTS

Intellectual property rights (IPRs); Biosafety concerns and regulations regarding GM crops

10.1. INTELLECTUAL PROPERTY RIGHTS (IPRs)

Intellectual property rights are rights of ownership or monopoly to an individual or a group or organization over creations developed by them through intellectual inputs. The right is given by a legal government agency for a stipulated time. This right prevents the use of such creations by others without permission. In this way the exclusive right of, for instance, an inventor over his or her invention is protected.

There are two types of IPRs, the Idustrial Property Rights and the Artisitic and Literary Property Rights with further sub categories.

Industrial property rights include, trade marks, industrial designs, geographical indications (GI), trade secrets and *patents*. Trade marks include distinct names and logos which identify a product. The right to industrial design covers the distinct design of the product involving colour, shape and other features. A common example of distinct trade marks and designs are the different types of mobikes available in the market, having different names, symbols and designs. GI norms have been developed to protect the quality and distinctive features of products originating from a particular geographical region. Some relevant examples are the wines from different vineyards and tea from a particular region as from Darjeeling in North Bengal. Trade secrets protect the rights on, for example, protocols of making a product for commercial purposes, like medicines of different brands.

Patents constitute a very important category of industrial property rights. *Patent* rights once granted, give exclusive rights for use and acquiring benefits of the invention to the inventor or *patentee*. A third party can use a patented item only after obtaining authorization and license from the patentee. Patent rights are granted for creations or inventions which have three components, *novelty*, element of *invention* and potential of *utility* or industrial application.

Creations of art, literature, written expression of ideas like papers, books are protected by the provisions of *copyright,* an item within the purview of Artisitic and Literary Property Rights.

There are possibilities of inventive items which do not strictly come under the provisions of categories mentioned above. An example is a plant variety developed through the ingenuity of a plant breeder. For such cases, like also database protection, an alternate mode has been developed, the *Sui Generis* system.

IPRs have two major concerns. These rights certainly protect the priority of and benefits accruing to the inventor and therefore, constitute an incentive for inventions. These rights also protect the commercial interests of the inventor or the licensee. But, a major concern is also ensuring protection of the *interests of the public*. Another concern was the variation in IPR norms, specially patent rights, between different countries. In 1995, therefore, a consensus was developed to bring uniformity of provisions in IPRs accross countries. The consensus was enshrined in Trade related aspects of IPRs, the TRIPS agreement.

Patent Law in India

Patent laws come within the realms of Industrial Property Rights.

India is a member of World Trade Organization (WTO). Therefore, as per TRIPs agreement it was necessary to modify the existing provisions of the Indian patent law. The provisions enshrined in the Indian Patent Act of 1970 were therefore, amended finally by January 1, 2005, for ensuring uniformity across member countries and TRIPs compliance.

The basic guidelines with regard to uniformity in patent laws for WTO members were given in article 27 of the TRIPS agreement. Thus, as per article 27.1 the item for patenting needs to be novel, inventive and useful for industrial application. *Novelty* implies that the item did not exist in the public domain prior to the filing of the application for patenting it. *Inventive* element refers to the item or idea not being obvious to the inventor prior to his development of the innovation. Industrial application implied the possibilities of the item being of large scale utility for the society. The provisions also ensured that there be no discrimination with regard to the region of origin of the technology or the field of technology of the patentable item. The TRIPS guidelines also indicated which items could not be patented. Thus, as per Article 27.2 any item /invention which disrupts *public order* or is detrimental to the *environment* cannot be patented. Similarly, therapeutic and surgical methods are disallowed for patenting by provisions of the Article 27.3. The original Indian Patent Act was modified within the purview of the above guidelines. The main features of the new Indian Patent Act, 2005 include the provisions which qualify which inventive products or processes can be patented and which items are to be excluded. Thus, as per Section 2 of the Act an invention is '*a new product or process involving an inventive step and capable of industrial application*'. The definition of inventive step needs some clarity. According to the same section of the Act an '*inventive step means a feature of an invention that involves technical advance as compared to the existing knowledge or having economic significance or both and that makes the invention not obvious to a person skilled in the art*'.

In this definition, the clause of nonobviousness is considered to mean that the invention is not obvious to the inventor person who is skilled in the art of the process, at the date and time the invention was actually made by him or her. For clarity about what can be patented, Section 3 lists the items which have to be excluded from patenting, that is, what are the items (products and processes) that cannot be patented. Some of the items in the list include: (i) any item that can upset 'public order', be detrimental to human, animal or plant life and environment, (ii) be a mixture of substances exhibiting combined properties of the components, (iii) medical, surgical, therapeutic and other treatments which cure diseases of humans and animals, (iv) plants, plant parts as seeds, animals, but not microbes and *essentially biological processes* for production and propagation of plants and animals, (v) inventions which constitute traditional knowledge or known properties of traditionally known substance/s. There are other items (vi) an agricultural or horticultural method. These are the main items of the list, in the context of biotechnological items for patenting.

Article 4 of TRIPS and section 3 of Indian Patent Act enumerate the items excluded for being essentially biological processes and which items are not. Thus, although processes as tranfection and transduction are excluded from patenting, processes which use some base sequences or patented cloning technique with permission can be patented. Similarly, a conventional plant breeding technique cannot be patented, but, processes involving human intervention as, enhancement of branching by some technique of pruning or growth increase by a compost mix can be patented.

Once a patent is granted, as per section 53 of the Indian Patent Act, in compliance with TRIPS, the patent rights are valid for 20 years from the date of application. Use of a patented process or product without permission from the patentee is considered illegal. Examples of patenting would include novel GMOs, cloning and expression vectors, microarrays and even important synthetic sequences.

Copyright

This priviledge comes under the category of Artisitic and Literary Property. A copyright assures ownership to an artistic or literary creation of a person. An author of an original paper has copyright on his/her writings and the original figures drawn. Copying by others without permission amounts to infringement of copyright and is punishable. In biotechnology, the items for copyrighting can be some original protocols of experimental procedures, designed sequence programmes and other text material incorporating an idea (but not the idea itself).

The *SUI GENERIS* System

In the TRIPS agreement plants and animals cannot be patented. However, microorganisms and methods of their reproduction can be patented. Because, plants and animals cannot be patented, TRIPS provides for an alternate possibility of protecting new plant varieties if needed. The alternate system is the *Sui Generis* system which member states may use to protect rights on a novel genetically modified plant variety developed. On the basis of the *Sui Generis* clause, Government of India enacted the 'Protection of Plant Varieties and Farmer's Rights Act, 2001' (PPVFR Act 2001). As per this act a *new plant variety* can be registered if it is novel, Distinct, Uniform and Stable (subject to DUS test). *Farmer's right* gives the farmers the right to save seeds, and improve and evolve plant varieties as individuals or communities.

It is necessary to mention that, it is the legal battle of an Indian scientist in US, Dr. Ananda Chakraborty, from 1970s to '80s, that paved the way for possibilities of patenting GMOs. Those were the days of environmental threats due to oil spills from tankers in seas and oceans. Dr. Chakraborty, a scientist in General Electrics labs, developed a GM bacterium of *Pseudomonas sp,* capable of digesting petroleum oil. This oil metabolizing bacterium had the potential of clearing oil spills if used under controlled conditions. In 1972 he had applied for patenting of his protocols and the GM bacterium by him. The US patent office allowed patenting of the protocols but rejected the application for patenting of a living organism, the GM bacterium. Then ensued a long legal battle and finally in the famous Diamond *vs* Chakraborty case, the US Supreme Court, allowed patenting of the GM oil digesting bacterium. Although, this microbe has not been employed for clearing oil spills for fear of its uncontrolled spread and consequent possible damage to global oil reserves, the judgement was pathbreaking in allowing patenting of GMOs. The court had decreed in this, Diamond *vs* Chakraborty case '*... anything under the sun made by man is patentable, provided there is novelty, inventiveness and industrial application..*'

Another landmark jugdement was given by the Calcutta High Court in the Dimminaco A.G. *vs* Controller of Patents and Design, in 1999, in connection with the process of making Bursitis vaccine for poultry. The court maintained that the process for making the vaccine product was *inventive* and the product although containing a *living microbe* had *vendibility* or marketability like any other manufactured commodity.

As mentioned above, although plants and animals cannot be patented, microbes can be patented. However, microbial cultures to be patented have to be deposited with some authority for checks whenever necessary. This is as per the Budapest Treaty of 1980, of which India became a signatory in 2001. India established the MTCC (Microbial Type Culture Collection) at the Institute of Microbial Technology (IMTECH) Chandigarh in 2002.

In India, microbial cultures for purposes of patenting have to be deposited at this centre of International Depository Authority.

10.2. BIOSAFETY CONCERNS AND REGULATIONS REGARDING GM CROPS

The key feature of transgenic crops is the vector-transgene recombinant DNA. In all GMOs produced by genetic engineering, it is the vector DNA that enables a transgene to move across or 'jump' across species. Recombinant DNA technology or Genetic Engineering is a quantum leap in technology. On the one hand, the technology is of great benefit and on the other, it does raise concerns. Scientists, right from the beginning have been concerned about the effect of the transgene in an alien host, the stability of the transgene in an alien ambience, the overall impact of the GMOs and specially GM crops on other organisms including human health and effects if any on the environment. The scientists practising recombinant DNA technology, the genetic engineers, have been and are actively engaged in improving the technology to make the GM crops safe. There are consistent efforts in this direction, such as, ensuring targeted insertion rather than random insertion, replacing antibiotic resistance markers with safe markers, production of marker free transgenics, use of inducible and tissue specific promoters, gene containment efforts including production of male sterile transgenics, transplastomic plants and even production of *cisgenic* plants with plant based regulatory elements. Such efforts can bear fruits only through concerted endeavour of adequately informed government authorites, private seed producers and the public. *The main thrust has to be ensuring, like in any other technology, the incorporation of proven technical advances for safety, into the commercial GM crop production protocols.* In such matters, ways must be found to even overcome hurdles posed by patenting norms.

As recombinant DNA science originated mainly in the United States, the US National Academy of Science deliberated on this issue and recommended some precautionary steps in early 1970s. The National Institute of Health, USA, in 1974, constituted a Recombinant Advisory Committee to study this issue and make recommendations for the way ahead. A historic conference of scientists was also held in 1975 at Asilomar, California to discuss the issues involved. The recommendations of the conference emphasized on the development of safety measures, assessment of risks of the protocols and materials used including safe bacteria and plasmids.

These initial deliberations have been followed by meetings of scientists, concerned authorities and civil society members in US, Europe, Japan and other countries, for developing biosafety measures in the production, release and use of GMOs.

In the case of genetic engineering of plants two situations arise: (a) production and release of GM crops/Transgenic plants, (b) use of plant cells for the production of biomaterials of interest. In the case of transgenic plants for use in agriculture and horticulture, the trangene-vector recombinant DNA is *uncontained*, as, it can spread in the environment through pollen dispersal. In the case of genetically engineered biomaterials *e.g.*, proteinaceous vaccines, the recombinant DNA, the vaccine geneDNA-vectorDNA complex, remains confined in the laboratories and is thus, *contained*. Therefore, in the case of GM crops, biosafety measures assume much greater significance. In the following sections biosafety protocols mainly related to GM crops have been discussed.

The objectives of biosafety regulations are to prevent as far as possible adverse effects of the release of GM crops on human health, animal health, wild and agro-biodiversity and the environment. *At the same time, it must encourage research and not be stifling.*

The main biosafety measure is Risk Assessment followed by introduction of measures for Risk Prevention and / or Risk Mitigation. Risk of any material or organism is a combination of its potential hazard and the frequency of exposure. In the case of transgenic crops, a transgene encoding a pesticide as Bt toxin, poses greater hazard than a transgene giving rise to nutrient factor like a vitamin A precursor. Similarly, the frequency of exposure is much higher in a cross pollinated crop than a self pollinated crop. Risk assessment involves the assessment of possible hazards. Hazards, as mentioned above, are mainly adverse effects on human health, animal (specially domestic) health and the environment. The assessment regimes may differ for different types of GM crops. Thus, the items and parameters for assessment will differ for transgenic cash crops and transgenic edible crops. The modules of assessment will vary for, for instance, higher nutrition edible crops as Golden rice, pesticide resistant cash crops as Bt cotton and pesticide resistant edible crops as Bt brinjal.

Transgenic plant production is evaluated in all the aspects of production protocol like, the use of promoters, the transgene of interest, the markers, regulators used and specially the expression quality and levels of the protein product and other features. Similarly, thorough post production assessment , as far as possible is also done. The pre-production assessments mainly have to be of: (a) the technology of gene delivery used, (b) the properties and characteristics of the vector components, like the promoters and markers, (c) the efficacy of transgene expression, (d) unintended effects of transgene insertion on host metabolism and other risks which may impair cellular processes. Some tests for monitoring transgene function have been mentioned in chapter 8.

After transfer of the GM crop to the field, besides agronomic performance, the assessments have to be mainly of the possible transfer of the transgene (horizontal gene transfer, HGT) to similar or other crops in the neighbourhood, (b) HGT to wild relatives, like weeds or wild specis in nearby forests, (c) impact on the health of domestic animals if the GM crop waste, specially Bt crop waste, is used as feed, (d) possibilities of development of resistant pests in response to pest resistant transgenic crops, (e) health risks to humans as, allergies, toxicity and other possibilities.

On the basis of the findings, measures for risk prevention and or risk mitigation are recommended. For instance, for Bt crops like Bt cotton there is recommendation for the use of rows of non transgenic plants of the same variety as 'refuge' plants. This is done to reduce the selection pressure of Bt toxin and helps in delaying the appearance of Bt toxin resistant boll worms. Similarly, there has been concern regarding the use of antibiotic resistance marker genes. Scientists felt that there could be possibilities of transfer of such genes to the gut microflora. Although an FDA (Food and Drug Adminstration, USA) study had shown insignificant chances of transfer of, for instance, the kanamycin resistance gene, FAO/WHO recommended avoiding the use of such markers.

Various international organizations, from 1990 onwards, have tried to develop a common framework or consensus for the safety assessment of specially GM foods. The organizations involved are FAO (Food and Agriculture Organization of the United Nation), WHO (World Health Organization), ILSI (International Life Sciences Institute), IFBC (International Food Biotechnology Council), OECD and specially the Codex Alimentarius Commission. A major safety assessment protocol consists of the determination of *Substantial Equivalence* between an edible transgenic plant and its non transgenic counterpart. The comparison is done with parameters as phenotype, biochemical composition and nutrition level. As plant specific differences in response to a transgene may exist, the OECD Task Force on Novel Foods and Feed initiated data compilation of different crops, the Consensus Documents, for comparative assessments.

Substantial equivalence studies are accompanied by parallel safety assessment studies. The safety assessments include investigations on the functioning of the transgene in the host, expression levels of the encoded protein, its activity and turn over; possible flow of the transgene and other genes inserted to other organisms; tests on potential allergenicity and toxicity of the transgene product; investigations on unintended effects.

Various protocols for testing possible allergenicity of the transgene encoded protein exist. There are difficulties of conducting targeted serum screening with human samples prone to allergies. Therefore, one of the favoured tests consists of finding the possible existence of sequence homologies between the transgene product and known allergenic proteins. This method, recommended by the FAO/

WHO Codex alimentarius Task Force on Foods Derived from Biotechnology, utilizes available data bases on sequences of allergenic proteins. A homology of 8 plus contiguous amino acids between the transgene product and any allergen is of concern and the GM product is subjected to further safety tests.

Several protocols exist for testing the toxicity of the protein/s encoded by the transgene/s. One of the tests is, like the allergenicity test, sequence comparison with known protein toxins, through data bases. However, main tests are done on experimental rats and mice. The main test is a prolonged test of 90 days feeding, the sub chronic test. The experimental animals are fed with rations mixed with known amounts of the transgenic protein and clinical signs recorded. The FAO/WHO Consultation in 2000 had recommended this as the initial test. If the results indicate something adverse, then other tests as for carcinogenicity, neurotoxicity, immunotoxicity and others can be conducted.

Besides, the above possibilities some unintended effects of transgene insertion can happen. It must be remembered that at present, in the commercially released GM crops, the transgene insertion is random. If the transgene gets inserted in a gene which is functional in normal growth, the resulting plantlet will be abnormal or will not survive. Therefore, the primary transgenic transformants which grow successfully, must have the transgene insertion in the noncoding region which does not impact normal growth. However, in transgenics for higher nutrition quality, as, Golden rice, unintended effects have been observed. In Golden rice, besides the intended product β-carotene, unintended chemicals as lutein and zeaxanthine are also formed. Therefore, to check out possibilities of any unintended effects, besides agronomic tests, comparative molecular profiling has been strongly recommended. The studies would include comparison of specially transcriptome, proteome and metabolic profiles of the parent and transgenic lines of the same plant species. For such studies, profile data bases of various plants are being constructed.

The pre-release or pre-market assessments are attempts to ensure biosafety in the short term. However, possibilities of long term adverse effects exist. Therefore, there are also recommendations for post market surveillance (PMS). As described above, several international and national organisations have developed and recommended safety norms, covering risk assessment, risk prevention and risk mitigation for transgenic crops. However, what is important is the implementation of these norms by different countries, growing or planning to introduce GM crops. Obviously, in each country or group of countries, the norms have to be adapted as per local needs and implemented through regulations developed by competent government authorities and expert bodies. The regulations mainly include, (a) items of pre and post production safety assessments to be conducted by producers of GM crops before commercial release, (b) monitoring of whether the seed producers are adhering to the norms or not, by competent bodies and or authorities, (c) pre-release evaluation of the safety assessment reports of the producers by competent bodies and or authorities, (d) provision for feedback from the public, mainly through civil society organizations, and (e) if possible post market surveillance. In a country like India, with wide ethnic diversity, involvement of each state in proposing location specific inputs for regulatory mechanisms also assume importance. Such regulatory mechanisms are essential as sooner or later, every country may grow GM crops or consume foods based on GM crops. Most regulatory systems have been derived from the Carte gena Biosafety Protocol.

In the United States, pre-production and pre-release regulation of, transgenic crops, in general, is by the USDA (US Department of Agriculture), the pest resistant crops, as, Bt crops by the EPA (Environment Protection Agency) and food safety of GM crops by FDA (Food and Drug Adminstration).

In the European Union, GM crop and GM food production, release and marketing are regulated by European Commission directives. These directives involve, monitoring of the processes at all stages and public feedback, both before release and marketing of the GM products.

In India, the Ministry of Environment and Forests (MoEF), Government of India, framed the initial regulations with regard to GMOs in 1989, under the provisions of the Environment Protection Act, promulgated in1986. Further, The rules of works related to Recombinant DNA, were framed by the Recombinant DNA Advisory Committee of the Department of Biotechnology in 1994. The researches on and commercial applications of GMOs, including transgenic plants is monitored by an apex body of the Government of India, the Genetic Engineering Approval Committee (GEAC). Various bodies at different levels, like the Institutional Biosafety Committee of (IBC) of institutions involved in R-DNA work, District Level Bodies, State Biotechnology Coordination Committees interact with the central Review Committee for Genetic Manipulation (RCGM) for overseeing biosafety aspects of GMOs, within the frameworks set by the GEAC.

An issue of interest is the *Precautionary Principle.* This principle envisages preventive measures for even remotely possible risks. Biological organisms in the context of a possible risk present some problems. Unlike an inanimate chemical drug or machine it is difficult to assess and predict both, the *likelihood* of a risk happening and the *magnitude* of the hazard if it happens. Imagine, hypothetically, the case of a transposon occurring in the chromosome, containing the p53 gene in humans. The p53 gene product checks a proliferative cell before division. If the cell is defective it induces apoptosis or cell death to prevent proliferation of the defective cell. If, however, the p53 gene gets mutated, the altered protein product loses the ability of checking and then the defective cells proliferate. Certain forms of cancer have been correlated with such mutated p53 genes. Now, coming to our hypothetical situation, the transposon mentioned may have a likelihood of jumping to another region of the same chromosome and the frequency of such jump/s can vary with the state of the cell. Further, it may jump to a noncoding region of the genome or into a gene, the mutation of which does not alter much, the metabolic state of the cell and thus such a jump does not affect the cell. But, the transposon can also jump and insert into the p53 gene, mutating it and resulting in induction of oncogeny and cancer. This example is an attempt to indicate that risks in biological organisms are uncertain and not calculable. Only probabilities exist. Therefore, many scientists and organizations promote the use of the precautionary principle. However, what must be borne in mind is that there should neither be under regulation nor over regulation.

In India, an ignored aspect of GM crops is the lack of concern for the VHY crop lines (very high yielding) being used as acceptors of the transgene, for commercial release. Further, instead of creating homozygous varieties for release, as in the US, the Bt cotton released in the market are hybrids with uncertain yields. It is necessary to promote location specific lines of good yield (and not very high yield) in the second step of GM crop production as acceptors of transgene from the primary transformant.

11

Timeline of Major Discoveries in Cell and Molecular Biology

SUMMARY OF CONTENTS

Historical perspective of discoveries from 1660s to date (Timeline, References and Links, Suggested reading)

Given hereunder is a timeline capsule of the main milestones in the history of cell and molecular biology.

TIMELINE OF DISCOVERIES AND INNOVATIONS

1665 Discovery of cells (cellular spaces in cork sections) by Robert Hooke in England[25].

1660s Discovery of unicellular organisms, *animalcules*, and bacteria by Anton van Leeuwenhoek in Holland[3].

1838 Description of living plant cells by Matthias Schleiden of Germany[56].

1839 Description of living animal cells by Theodor Schwann of Germany. Both proposed the *cell theory* – cell is the structural unit of life[56].

1855 Cells arise from pre-existing cells proposed by Rudolf Virchow in Germany[64].

1866 Theory of inheritance of traits through 'units of heredity' which can segregate and show random assortment during inheritance from parents to progeny given by Gregor Mendel of Austria[38] (paper, rediscoverd in 1900).

1871 Nucleic acid extracted from WBC of pus and termed *nuclein* by Friedrich Miescher, from Ernst Hoppe-Seyler's laboratory, Tubingen, Germany[41,42]. In 1889, nuclein was renamed as *Nucleic Acid* by Miescher's student Richard Altmann in Switzerland[2].

1882 Threadlike 'coloured bodies' discovered by Walther Flemming of Germany, named later as *chromosomes*[18].

1887 Reduction division, halving of chromosome number during gamete formation, proposed by the German biologist August Weissman[67].

1888 Thread like 'coloured bodies' described by Flemming were named as *chromosomes* by W. Waldeyer of Germany[65].

1897 Enzymes as biological catalysts discovered by Eduard Buchner in France[11].

1902 First attempts for plant tissue culture by Gotleib Haberlandt[22].

1903 Discovery of homologous chromosomes, chromosome behaviour in mitosis and meiosis by Walter Sutton in USA[60].

1908 The concept that some diseases are genetic being 'inborn errors of metabolism was proposed by Archibald Garrod of UK, on the basis of studies of the alcaptonuria disease. This was like a precursor of the 'one gene one enzyme' hypothesis[21].

1909 The 'units of heredity' were named as *genes* by Wilhelm Johannsen of Denmark[29].

1909–1915 Through mutation and other studies on the fruit fly *Drosophila,* Thomas Hunt Morgan and his students showed mainly that, genes are located on chromosomes; each chromosome is a linkage group of genes; during meiosis, crossing over or recombination takes place between homologous chromosomes[44]; Alfred Sturtevant in 1911, from Morgan's laboratory initiated construction of chromosome maps showing location of genes.

1926 Enzymes shown to be proteins by James B Sumner of USA[59].

1928 Transformation of *avirulent Pneumococcus* bacteria to *virulent* form by mixing with a suspension of heat killed virulent bacteria, by Frederick Griffith in UK, provided a major clue regarding the nature of the chemical basis of heredity[20].

1933 Invention of Electron microscope by Ernst A. F. Ruska[52].

1944 The transforming principle in Griffith's experiment was shown to be DNA by Oswald Avery, Colin MacLeod and Maclyn McCarty of USA[5].

1950s Mobile elements in chromosomes, the 'jumping genes' or transposons were discovered by Barbara McClintock, NY, USA[37].

1941 'One gene one enzyme' hypothesis proposed on the basis studies on *Neurospora* mutants by George Beadle and Edward Tatum, California, USA[7].

1950s Cell fractionation and ultrastructure studies of Albert Claude[13], Christian de Duve[17] and George E Palade[49], revealed the roles of organelles as ribosomes, ER, lysosomes.

1951 Secondary structure of protein, α-helix, elucidated by Linus Pauling and R.B. Corey[50].

1952 DNA as the chemical basis of heredity was conclusively proven with the demonstration by Alfred Hershey and Martha Chase, in USA, that bacteriophages transfect by inserting their nucleic acid core only, into bacteria followed by production of progeny virions[23].

1953 DNA double helical structure proposed by James Watson and Francis Crick, from Cambridge, UK[66] on the basis of X-ray diffraction pictures of DNA prepared by Rosalind Franklin and Maurice Wilkins and DNA base ratios reported by Erwin Chargaff.

1958 DNA polymerase enzyme discovered by Arthur Kornberg and co-workers[8] and RNA polymerase by Severo Ochoa and co workers in USA[48].

1958 DNA replication is semi conservative was demonstrated by Matthew Messelson and Frank Stahl[40].

1959 Amino acid sequence of the protein, bovine insulin, elucidated by Frederich Sanger[53]. Myoglobin structure elucidated[30].

1961 Messenger RNA discovered by Francois Jacob and Jacques Monod of France[28], Sydney Brenner of UK and Matthew Messelson of USA[10].

1960s The bacterial *operon model* proposed by Francois Jacob, Jacques Monod and co-workers France[27]. Later they discovered also allosteric enzymes. So, Jacob and Monod are credited with three discoveries, operon model, mRNA and allosteric enzymes[43].

1961–1964 Triplet codons for amino acids elucidated with messenger binding assay (adding synthetic trinucleotides one at a time to a mixture of ribosome, amino acyl-tRNAs etc., and detecting the specific aa-tRNA that gets bound to the ribosome trinucleotide complex) by Marshall Nirenberg, J H Matthei, and Philip Leder, of USA[33,35] and Francis Crick of UK[15].

1960s Synthetic triplet codes constructed by Har Govind Khorana in USA[31].

1965 First sequence of yeast alanine transfer RNA reported by Robert Holley of USA[24].

1965–1970 Discovery of action and applications of restriction endonucleases by W. Arber[4], H.O. Smith[57] and Daniel Nathans[16], in USA and Europe.

1970 Discovery of reverse transcriptase by Howard M Temin[61], David Baltimore[6] and Renato Dulbecco.

1972–1974 Recombinant DNA technology, (mainly making of plasmid vector-goi recombinant and cloning of alien gene in *E. coli*) pioneered by Paul Berg, Stanley Cohen, Herbert Boyer, Janet Mertz, Ronal Davis and their co-workers in USA[14,39,26].

1975 Development of the Southern Blotting method for DNA separation and detection by E.M. Southern in UK[58]. Similar blotting for RNA, Northern Blot, was developed by J C Alwine and colleagues in USA[1] and for proteins, Western Blot, by Towbin and coworkers in USA[62].

1977 DNA sequencing methods developed by A M Maxam and W Gilbert[36] as also F Sanger[54,55].

1977 The plasmid pBR322 developed by F Bolivar and RL Rodriguez[9].

1979 Chemical synthesis of gene by Har Govind Khorana[32].

1983 Polymerase Chain Reaction (PCR) developed by Kary Mullis of USA[46].

1985–2005 Initiation and near completion of the Human genome Project.

1986 First field trials of GM crops done in USA and France.

1998 Discovery of RNA interference, RNAi by Andrew Z Fire and Craig C Mello[19].

2000 Discovery of RNA catalysts, the ribozymes by T R Cech[12].

2010 First bacterial genome synthesized at J Craig Venter Institute and introduced in nucleic acid free bacterial cells which multiplied and produced synthetic progeny, the *Synthia* bacteria, the world's first synthetic life form.

1980s to date In India, the dawn of the era of studies on genome organization, functional genomics, genome wide analysis (genome sequencing, marker assisted selection, QTLs) in various laboratories supported mainly by the Department of Biotechnology and the Indian Council of Agriculture Research[63].

REFERENCES AND LINKS

[1] Alwine J.C., Kemp, D.J. and Stark, G.R. (1977). Method for detection of specific RNAs in agarose gels by transfer to diazobenzyloxymethyl-paper and hybridization with DNA probes. *Proc. Natl. Acad. Sci. U.S.A.,* 74: 5350–54.

[2] Altman, R. (1889). *Anat. U. Physiol., Physiol. Abt.,* 524

[3] Anton van Leeuwenhoek, 1660s, Letters to The Royal Society, London, http://www.vanleeuwenhoek.com/Letters.htm

[4] Arber, W. (1965). Host controlled modification of bacteriophage. *Annu. Rev. Microbiol.,* 19: 365–378

[5] Avery, O.T., Macleod, C.M. and McCarty, M. (1944). Studies on the chemical nature of the substance inducing transformation of the Pneumococcal types: induction of transformation by a deoxyribonucleic acid fraction isolated from Pneumococcus type III. *J. Exp. Med.,* 6: 201–209.

[6] Baltimore, D. (1970): RNA-dependent DNA polymerase in virions of RNA tumour viruses. *Nature*, 226: 1209–1211.

[7] Beadle, G.W. and Tatum, E.L. (1941). Genetic control of biochemical reactions in *Neurospora. Proc. Natl. Acad. Sci.,* 27: 499–506

[8] Bessman, M.J., Lehman, I.R., Simms, E.S. and Kornberg, A. (1958). Enzymatic synthesis of deoxyribonucleic acid. II. General properties of the reaction. *J. Biol. Chem*., 233: 171–177.

[9] Bolivar, F., Rodriguez, R.I., Greene, P.J., Betlach, M.C., Hyneker, H.L., Boyer, H.W., Crosa, J.H. and Falkow, S. (1977). Construction and characterization of new cloning vehicles. II. A multipurpose cloning system. *Gene*, 2: 95–113.

[10] Brenner, S., Jacob, F. and Messelson, M. (1961). An unstable intermediate carrying information from genes to ribosomes for protein synthesis. *Nature*, 190: 576–581

[11] Buchner E.: http://en.wikipedia.org/wiki/Eduard_Buchner

[12] Cech, T.R. (2000). The ribosome is a ribozyme. *Science*, 289: 878–879.

[13] Claude, A. : http://en.wikipedia.org/wiki/Albert_Claude

[14] Cohen, S.N., Chang, A.C.Y., Boyer, H.W. and Helling, R.W. (1973). Construction of biologically functional bacterial plasmids *in vitro. Proc. Natl. Acad. Sci., USA,* 70: 3240–3244

[15] Crick F.H.C., Brenner, S., Barnett, L. and Watts-Tobin, R.J. (1961). General nature of the genetic code for proteins. *Nature.* 192: 1227–1232.

[16] Danna. K. and Nathans. D. (1971). Specific cleavage of simian virus 40 DNA by restriction endonuclease of *Hemophilus influenza. Proc. Natl. Acad. Sci., USA*, 68: 2913–2917.

[17] De Duve: http://en.wikipedia.org/wiki/Christian_de_Duve

[18] Flemming Walther: http://en.wikipedia.org/wiki/Walther_Flemming

[19] Fire, A., Xu, S., Montgomery, M., Kostas, S., Driver, S. and Mello, C. (1998). Potent and specific genetic interference by double-stranded RNA in *Caenorhabditis elegans. Nature*, 391: 806–811.

[20] Griffith, F. (1928). The significance of Pneumococcal types. *J. Hygiene*, 27: 113–159.

[21] Garrod, A.: http://en.wikipedia.org/wiki/Archibald_Garrod

[22] Haberland, G., http://agriinfo.in/default.aspx?page=topic&superid=3&topicid=1881

[23] Hershey, A.D. and Chase, M. (1952). Independent functions of viral protein and nucleic acid in growth of bacteriophage. *J. Gen. Physiol.,* 36; 39–56.

[24] Holley, R.W., Everett, G.A., Madison, J.T. and Zamir, A. (1965). Nucleotide sequences in the yeast alanine transfer ribonucleic acid. *J. Biol. Chem.,* 240 (5): 2122–2128.

[25] Hooke, R. (1665), Micrographia: Physiological descriptions of minute bodies made by magnifying glasses, Printed by Martyn and Alleftry for The Royal Society, London.

[26] Jackson, D.A., Symons, R.H. and Berg, P. (1972), Biochemical method for inserting new genetic information into DNA of Simian virus 40: Circular SV40 DNA molecules containing lambda phage genes and the galactose operon of *Escherichia coli. Proc. Natl. Acad. Sci. USA.,* 69: 2904–2909.

[27] Jacob, F., Perrin, D., Sanchez, C. and Monod, J. (1960). Operon: A group of genes with the expression coordinated by an operator. *Academie des Sciences,* 250(6): 1727–1729.

[28] Jacob, F. and Monod, J. (1961), Genetic regulatory mechanisms in the synthesis of proteins. *Journal of Molecular Biology*, 3: 318–356.

[29] Johannsen, W.: http://natureinstitute.org/pub/ic/ic14/gene.htm

[30] Kendrew, J.C. (1959). Structure and function in myoglobin and other proteins. *Federation Proceedings,* 18(2, Part 1): 740–51.

[31] Khorana, H.G. (1960). Synthesis of nucleotides, nucleotide coenzymes and polynucleotides. *Federation Proceedings*, 19: 931–941.

[32] Khorana, H.G. (1979). Total synthesis of a gene. *Science*, 203: 614–625

[33] Leder, P. and Nirenberg, M.W. (1964). RNA codewords and protein synthesis, III. On the nucleotide sequence of a cysteine and a leucine RNA codeword. *Proc. Natl. Acad. Sci., USA*, 52(6): 1521–1529.

[34] Lehman, I.R., Bessman, M.J., Simms, E.S., and Kornberg, A. (1958). Enzymatic synthesis of Deoxyribonucleic Acid. I. Preparation of substrates and partial purification of an enzyme from *Escherichia coli* , *J. Biol. Chem.,* 233, 163-170

[35] Matthai, H.G., Jones, O.W., Martin, R.G. and Nirenberg, M.H. (1962). Characteristics and composition of RNA coding units. *Proc. Natl. Acad. Sci., USA*, 48(4): 666–677.

[36] Maxam, A.M. and Gilbert, W. (1977). A new method of sequencing DNA. *Proc. Natl. Acad. Sci., USA.,* 74: 560–564.

[37] McClintock, B. (1956). Controlling elements and the gene, in Genetic Mechanisms: Structure and Function, Cold Spring Harbour Symposia on Quantitative Biology, Vol XXII.

[38] Mendel, G. (1865). Experiments in plant hybridization, http://www.mendelweb. org/Mendel.html

[39] Mertz, J.E. and Davis, R.W. (1972). Cleavage of DNA by R I restriction endonuclease generates cohesive ends. *Proc. Natl. Acad. Sci. USA*, 69: 3370–3374.

[40] Messelson, M and Stahl, F.W. (1958). The replication of DNA in *Escherichia coli. Proc. Natl. Acad. Sci., USA*, 44: 671–682

[41] Miescher, F. (1871). Ueber die chemische zusammensetzung der eiterzellen, *Medicinisch- Chemische Untersuchungen,* 4: 441–460.
[42] Friedrich Miescher Biography, www.dnaftb.org/15/bio.html.
[43] Monod, J., Changeux, J. and Jacob, F. (1963). Allosteric enzymes and cellular control systems. *Journal of Molecular Biology*, 6, 306-329
[44] Morgan, T. H., Sturtevant, A.H. and Bridges, C.B. (1919). Contributions to the genetics of *Drosophila melanogaster,* Published by the Carnegie Institution, Washington DC, USA
[45] Morrow J.F., Cohen, S.N., Chang, A.C.Y., Boyer, H.W., Goodman, H.M. and Helling, R.B. (1974). Replication and transcription of eukaryotic DNA in *Escherichia coli. Proc. Natl. Acad. Sci., USA*, 71: 1743–1747.
[46] Mullis, K., Faloona, F., Scharf, S., Saiki, R., Horn, G. and Ehrlich, H. (1986). Specific enzymatic amplification of DNA *in vitro*: The polymerase chain reaction. *Cold Spring Harbour Symposium in Quantitative Biology*, 51: 263–273.
[47] Nirenberg, M.W. and Matthai, H.J. (1961). The dependence of cell free protein synthesis in *E. coli* on naturally occurring or synthetic polyribonucleotides. *Proc. Natl. Acad. Sci.,* 47(10): 1588–1602 .
[48] Ochoa, S., http://en.wikipedia.org/wiki/Severo_Ochoa
[49] Palade, G.E.: http://en.wikipedia.org/wiki/George_Emil_Palade
[50] Pauling, L., Corey, R.B. and Branson, H.R. (1951). The structure of proteins: Two hydrogen bonded helical configurations of the polypeptide chain. *Proc. Natl. Acad. Sci.,* 37(4): 205–211.
[51] Perutz, M. (1962). Proteins and nucleic acids: Structure and Function, Elsevier, London.
[52] Ruska, E.: http://www.nobelprize.org/nobel_prizes/physics/laureates/1986/ruska-lecture.html
[53] Sanger, F. (1959). Chemistry of Insulin: determination of the structure of insulin opens the way to greater understanding of life processes. *Science,* 129 (3359): 1340–1344.
[54] Sanger, F., Nicklen, S. and Coulsen, A.R. (1977). DNA sequencing with chain terminating inhibitors. *Proc. Natl. Acad. Sci., USA*, 74: 5463–5467.
[55] Frederick Sanger Biography, http://www.nobelprize.org/nobel_prizes/chemistry/laureates/1980/sanger-bio.html.
[56] Schwann, T. and Schleyden, M.J. (1847). Microscopical researches into the accordance in the structure and growth of animals and plants, Sydenham Society Publication, London
[57] Smith, H.O. and Wilcox, K.W. (1970). A restriction enzyme from *Hemophilus influenzae* 1.Purification and general properties. *J. Mol. Biol.*, 51: 379–391.
[58] Southern, E. M. (1975). Detection of specific sequences among DNA fragments separated by gel electrophoresis. *J. Mol. Boil.,* 98: 503–517.
[59] Sumner, J.B. (1927). Text book of Biochemistry, Macmillan, NY
[60] Sutton, W.S. (1903). The chromosomes in heredity. *Biol. Bull.,* 4: 231–251.
[61] Temin, H.M. and Mizutani, S. (1970). RNA-dependent DNA polymerase in virions of *Rous sarcoma* virus. *Nature*, 226: 1211–1213.

[62] Towbin, H., Stahelin, T. and Godon, J. (1979). Electrophoretic transfer of proteins from polyacrylamide gels to nitrocellulose sheets: Procedure and some applications. *Proc. Natl. Acad. Sci. USA*, 76: 4350–4354.

[63] Tyagi, A.K. and Khurana, J.P. (2003). Plant molecular biology and biotechnology research in the post recombinant DNA era. *Adv. Biochem. Engin / Biotechnol.,* 84: 91–121.

[64] Virchow, R.L.K., http://en.wikipedia.org/wiki/Rudolf_Virchow

[65] Waldeyer, W.:http://en.wikipedia.org/wiki/Heinrich_Wilhelm_Gottfried_von_Waldeyer-Hartz.

[66] Watson, J.D. and Crick, F.H.C. (1953). A structure for deoxyribose nucleic acid, *Nature*, 171: 737–738.

[67] Weissman, A.: http://en.wikipedia.org/wiki/August_Weismann

SUGGESTED FURTHER READING

Acquaah, G. (2004). Understanding Biotechnology, Pearson Education Pvt. Ltd., Indian Branch, Delhi, India

Barnum, S.R. (2005). Biotechnology an Introduction, Thomson Brooks / Cole, USA

Brown, T.A. (2007). Genomes 3, Garland Science Publishing, NY, USA

Brown, T.A. (2010). Gene Cloning and DNA Analysis *an Introduction*, Wiley-Blackwell, Sussex, UK

Buchanan, B.B., Gruissem, W. and Jones, R.L. (2000). Biochemistry and Molecular Biology of Plants, American Society of Plant Physiologists, Rockville, Maryland, USA

Chopra, V.L., Santharam, S.P. and Sharma, R.P. *eds*., (2005), Biosafety of Transgenic Rice, National Academy of Agriculture Sciences, New Delhi, India

Clark, D.P. (2005). Molecular Biology Understanding the Genetic Revolution, Elsevier Inc., NY, USA

Dale, J.W. and von Schantz, M. (2007). From Genes to Genomes, John Wiley and Sons Ltd., Sussex, UK

Das, H.K. *ed*., (2010), Textbook of Biotechnology, Wiley India Pvt. Ltd., New Delhi, India

De Robertis, E.D.P. and De Robertis, E.M.F. (2001). Cell and Molecular Biology, Lippincott Williams and Wilkins, PA, USA

Gardner, E.J., Simmons, M.J. and Snustad, D.P. (1991). Principles of Genetics. John Wiley and Sons Inc., NY, USA

Glick, B.J. and Pasternak, J.J. (1998). Molecular Biotechnology: Principles and Applications of Recombinant *DNA*, American Society for Microbiology, Washington DC, USA

Grierson D. *ed*., (1991). Plant Genetic Engineering, Springer (India) Pvt. Ltd., New Delhi, India

Gupta, P.K. (2010). Plant Biotechnology, Rastogi Publications, Meerut, India

Karp, G., Cell and Molecular Biology: Concepts and Experiments, John Wiley and Sons Pvt. Ltd., NJ, USA

Klug, W.S. and Cummings, M.R. (2003). Concepts of Genetics, Dorling Kindersley (India) Pvt. Ltd., New Delhi, India

Lewin, B. (2008). Genes IX, Jones and Bartlett Publishers, Boston, USA
Lodge, J., Lund, P. and Minchin, S., Gene Cloning: Principles and Applications, Taylor and Francis, NY, USA
Micklos, D.A., Fryer, G.A. and Crotty, D.A. (2003). DNA Science, Cold Spring Harbour, NY, USA
Nelson, D.L. and Cox, M.M. (2000). Lehninger's Principles of Biochemistry, Macmillan / Worth Publishers, NY, USA
Primrose, S.B. and Twyman, R.M. (2006). Principles of Gene Manipulation and Genomics, Blackwell Publishing, MA, USA
Shah, M., Banerji, D., Vijayshankar, P.S. and Ambasta, P. (1998). India's Drylands: Tribal Societies and Development through Environmental Regeneration, Oxford University Press, New Delhi,
Sharma, J.R. (1994). Principles and Practice of Plant Breeding, Tata McGraw-Hill Publishing Company Ltd., New Delhi, India
Singh, B.D. (1993). Plant Breeding, Kalyani Publishers, Ludhiana, India
Singh, B.D. (2004). Biotechnology: Expanding Horizons, Kalyani Publishers, Ludhiana, India
Slater, A., Scott, N.W. and Fowler, M.R. (2008). Plant Biotechnology, Oxford University Press, NY, USA
Stewart Jr, C.N. *ed.*, (2008). Plant Biotechnology and Genetics, John Wiley and Sons, Inc., NJ, USA
Walker, J.M. and Rapley, R. *eds.* (2002). Molecular Biology and Biotechnology, Panima Publishing Corporation, New Delhi, India
Watson, J.D., Baker, T.A., Bell, S.P., Gann, A., Levine, M. and Losick, R. (2004). Molecular Biology of the Gene, Pearson Education Pte Ltd., Delhi, India
Watson, J.D., Myers, R.M., Caudy, A.A. and Witkowski, J.A. (2007). Recombinant DNA, W. H. Freeman and Company, NY, USA
West, E.S., Todd, W.R., Mason, H.S. and Van Bruggen, J.T. (1966). Biochemistry. Collier-Macmillan Canada Ltd., Toronto, Canada

SUBJECT INDEX

D

E

F

M

N

O

P

T

U

V

W

X

Y

Z